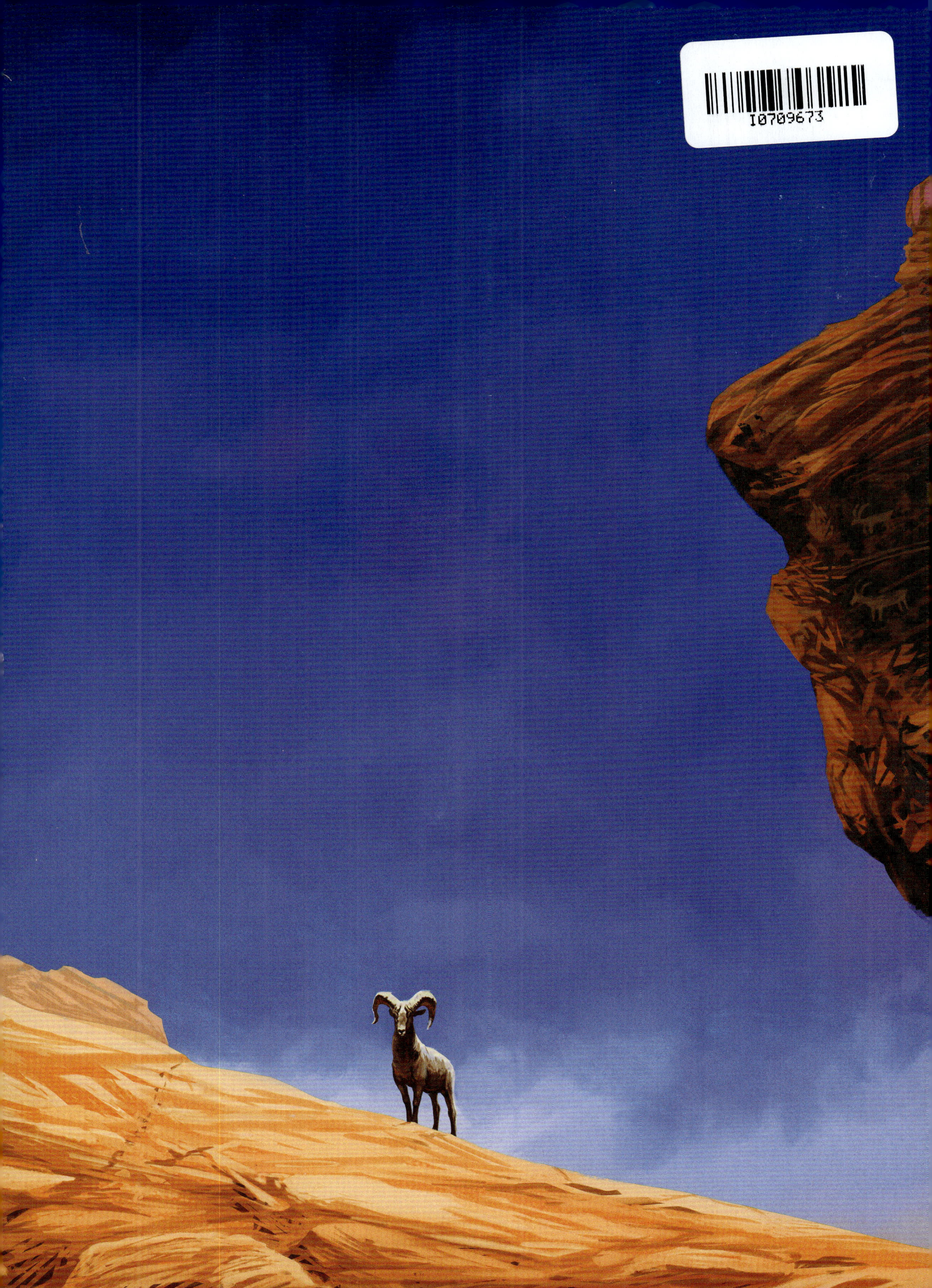

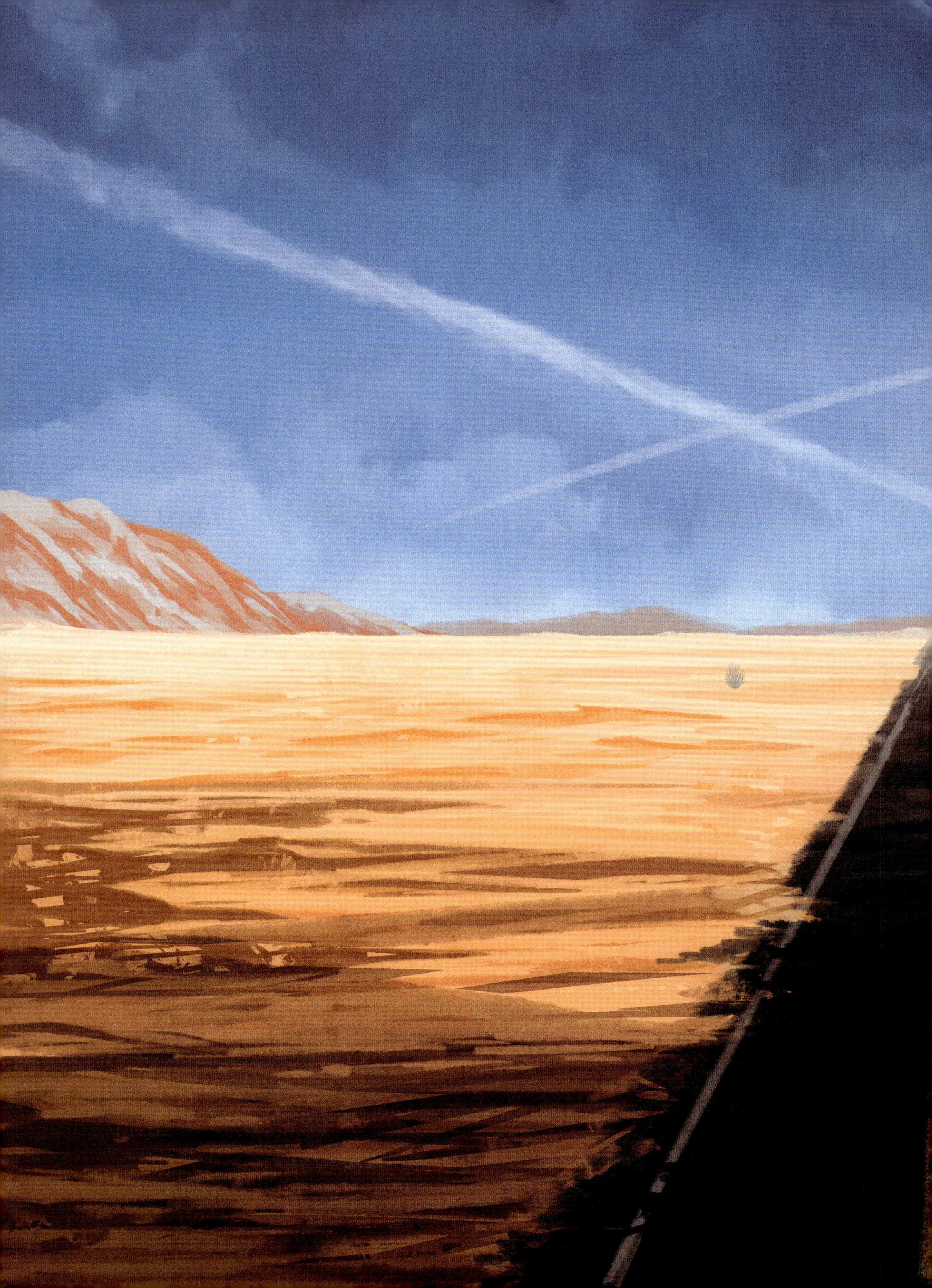

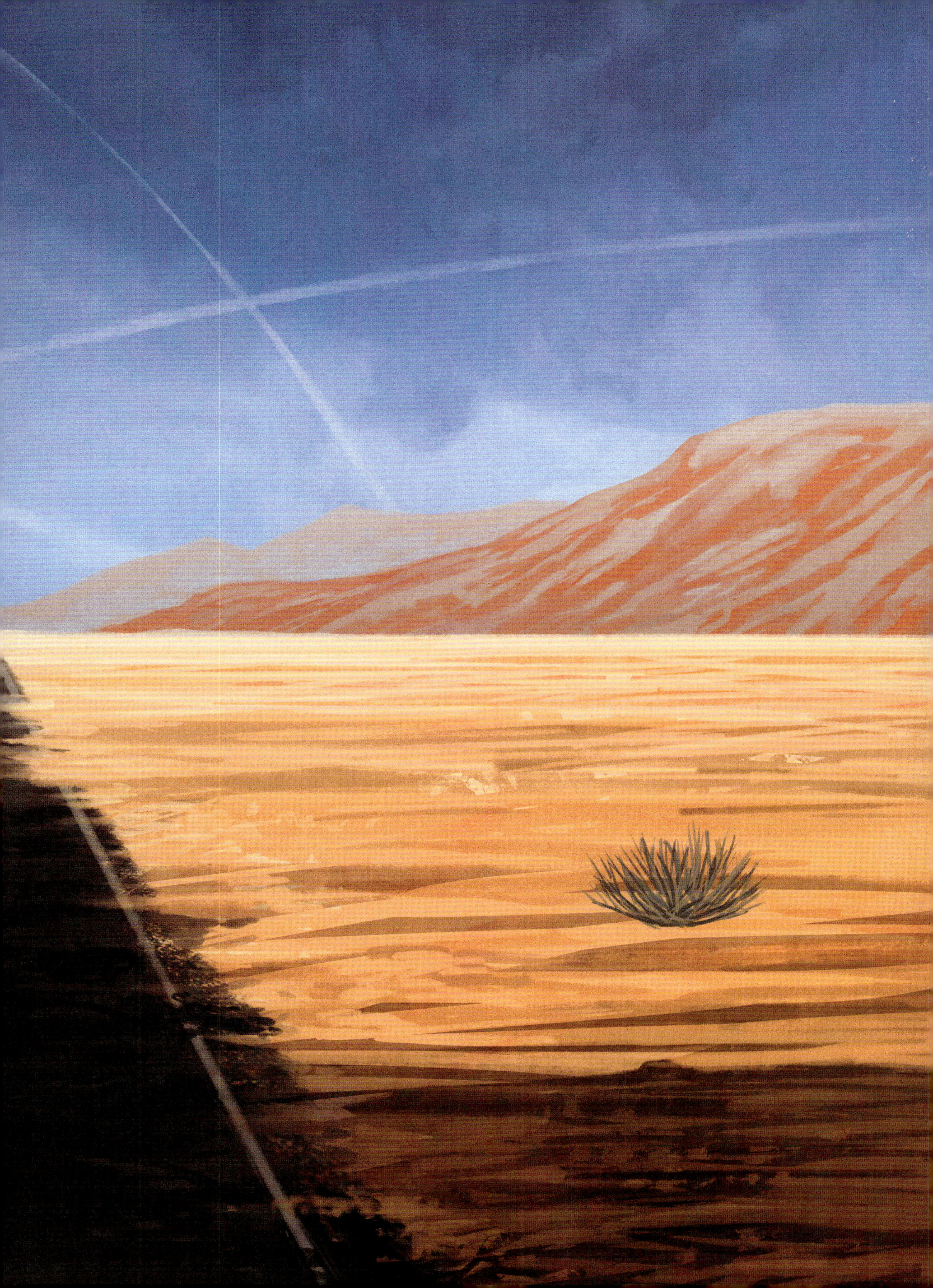

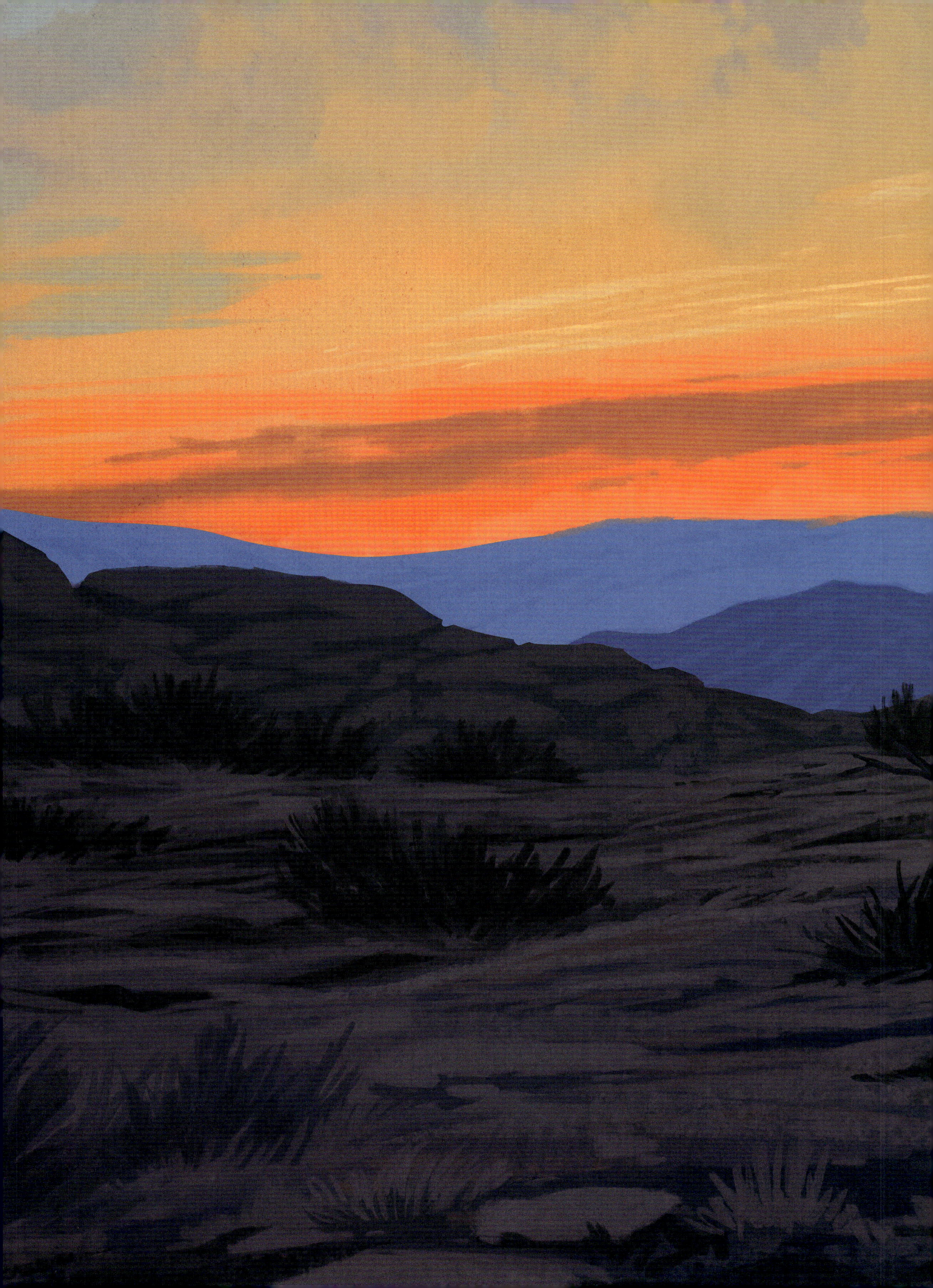

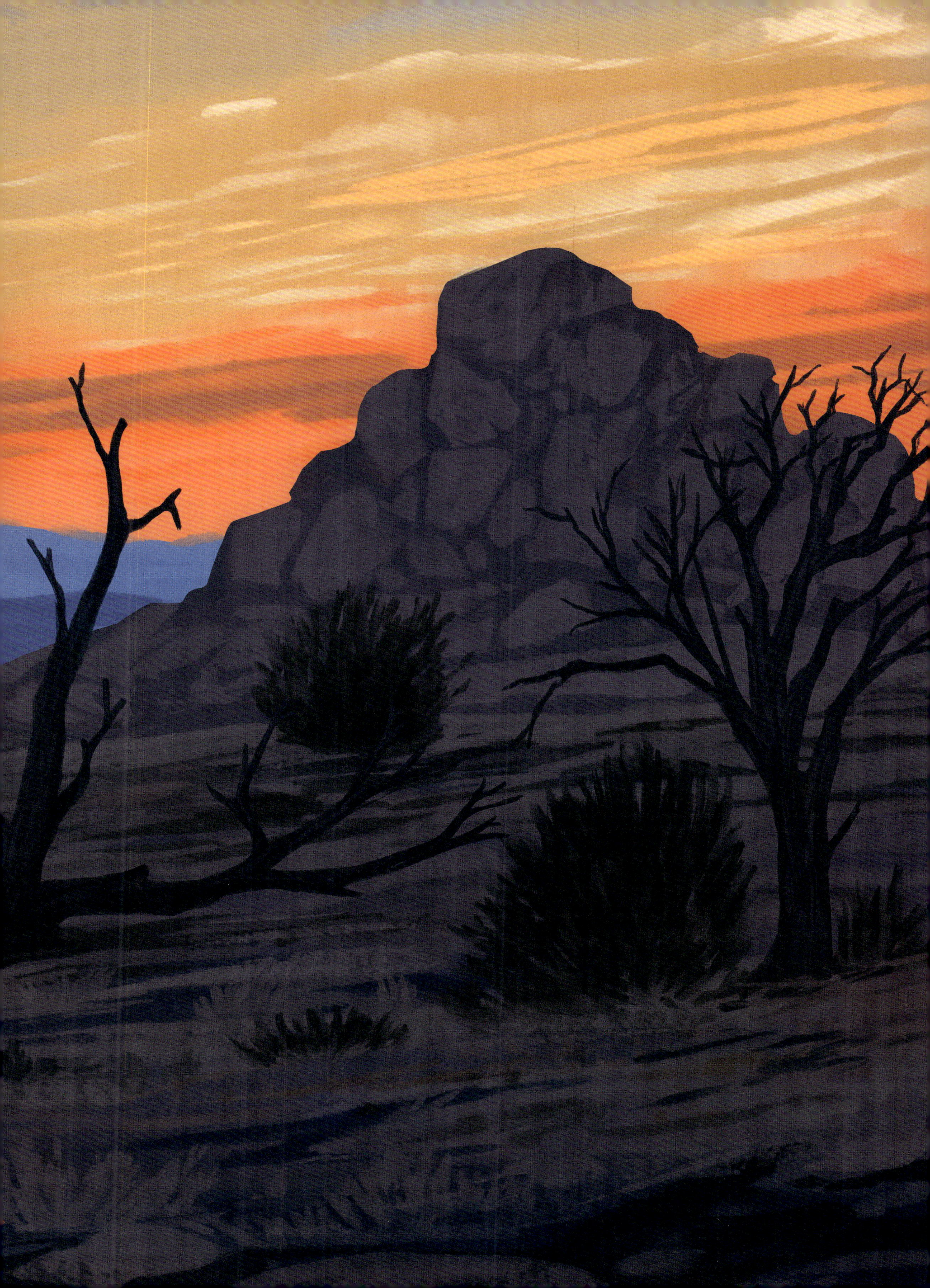

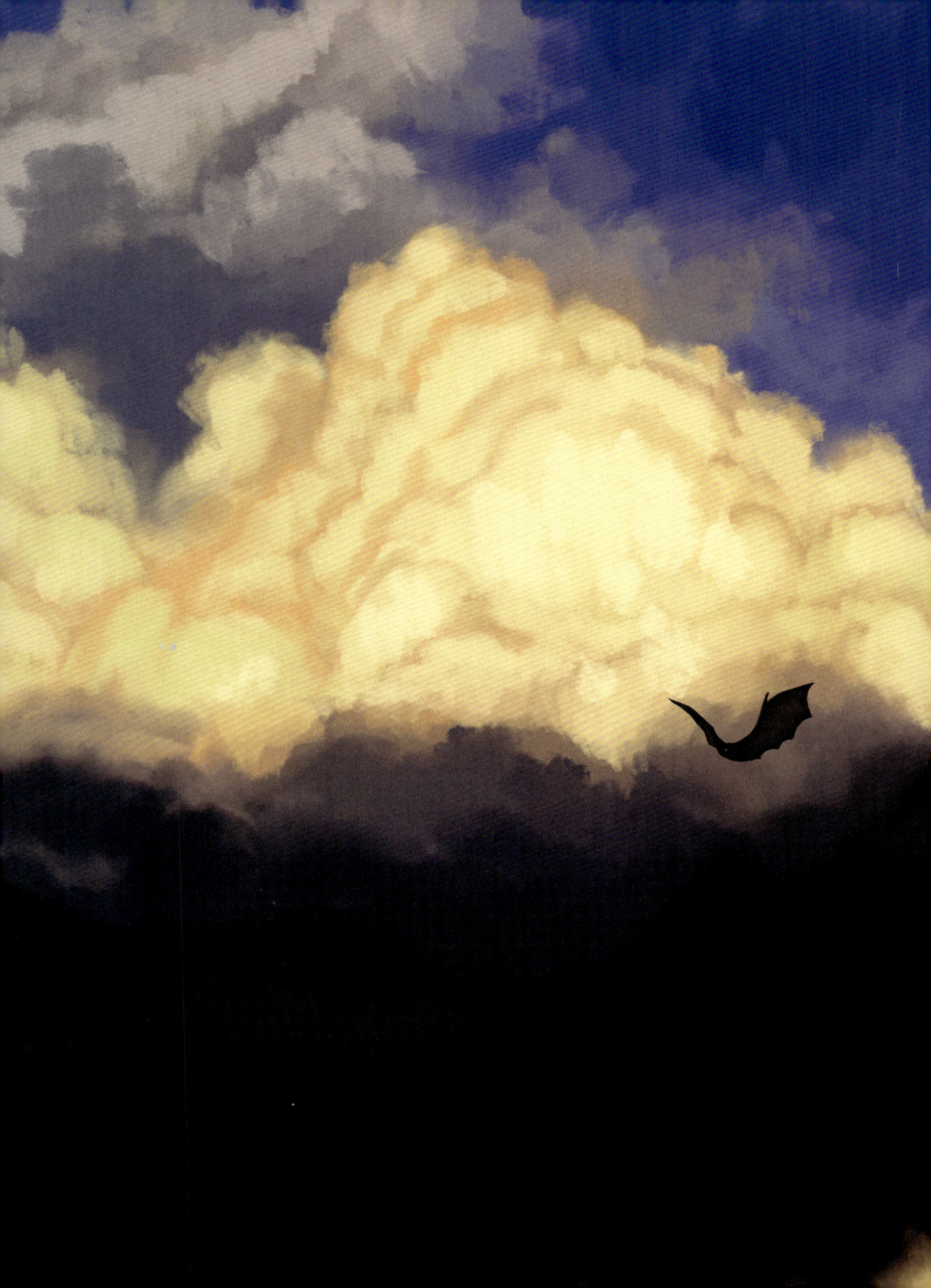

BARREN LANDS

WRITTEN AND ILLUSTRATED BY

BRYNN METHENEY

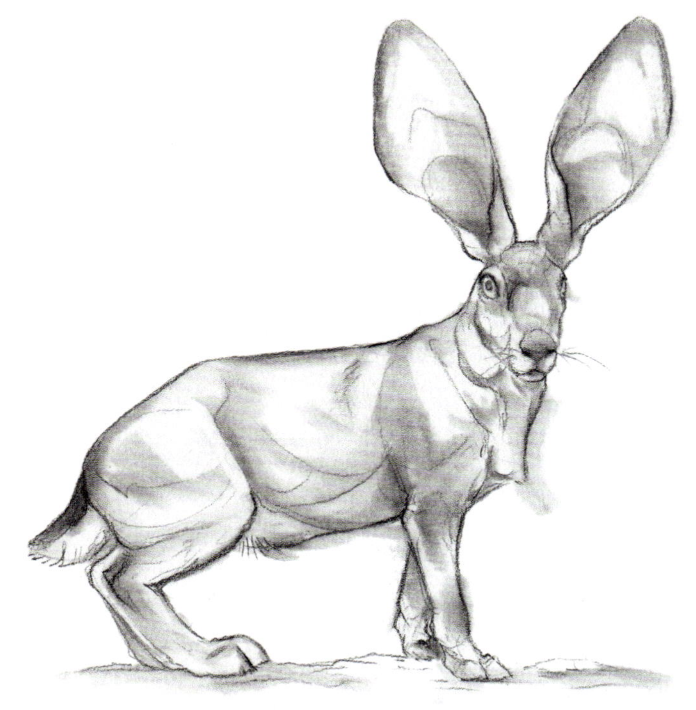

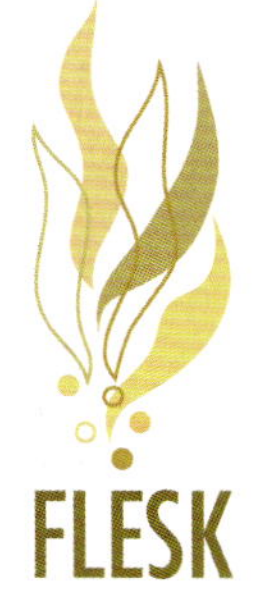

FLESK

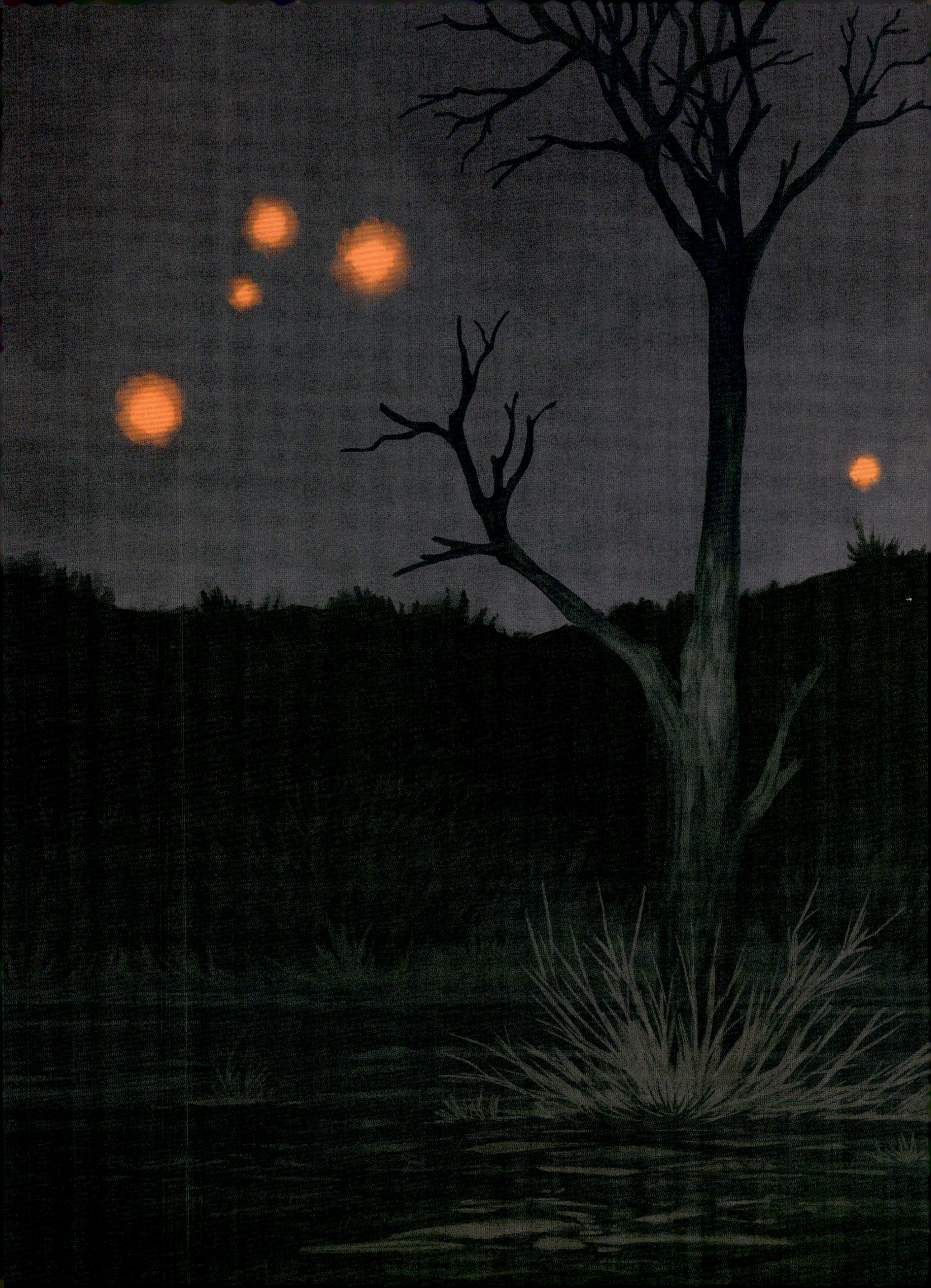

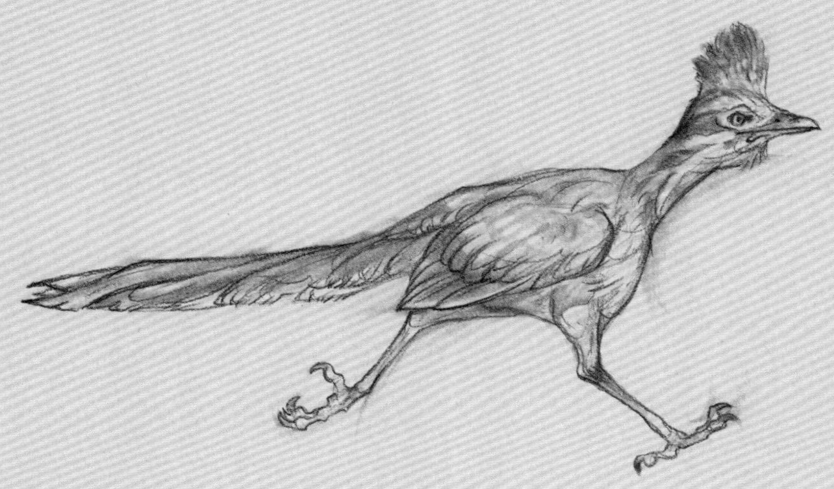

INTRODUCTION

Many folks think the desert is a desolate place. Most would say nothing lives here: A "barren wasteland" they call it. They couldn't be more misinformed.

In the North American Southwest, vast expanses of wild desert peppered with human sprawl and decaying ghost towns means that driving is a big part of daily life in these parts. And when you do a lot of driving, you usually find yourself traveling on some back road somewhere, looking into that endless horizon.

What they don't tell you is that if you do this enough, you begin to crave it. You see, driving on those lonely roads opens something. You begin to see things, and you notice the creatures and the plants and the beings that are out there.

You gain appreciation for the way the desert paints itself a million different colors.

You start to see lights.

Animals begin to appear next to your car, and you wonder if maybe you died out there somewhere and you returned to haunt the place.

This collection of stories and artworks strives to document the seasonal occurrences, natural beauty and strange phenomena of the North American Southwestern deserts that I've personally come across. I've included notes, observations, family stories and such.

In some instances, coordinates and locations are provided. You are encouraged to visit the sites, but please do so with cautious planning. Listed below each location are considerations to follow. Pay close attention. Not doing so could be dangerous.

Follow and respect local laws on Native lands. Remember, you're a visitor here. Reverence is required.

Certain locations have been omitted to preserve the natural and supernatural integrity of the places and things I've documented. I appreciate you not looking for them.

Only drive along established roads that are permitted for such things. When roads are flooded, turn around! Don't drown! Fuel up, and bring a couple gallons of water, a bandana and a paper map.

Stay safe, friend.

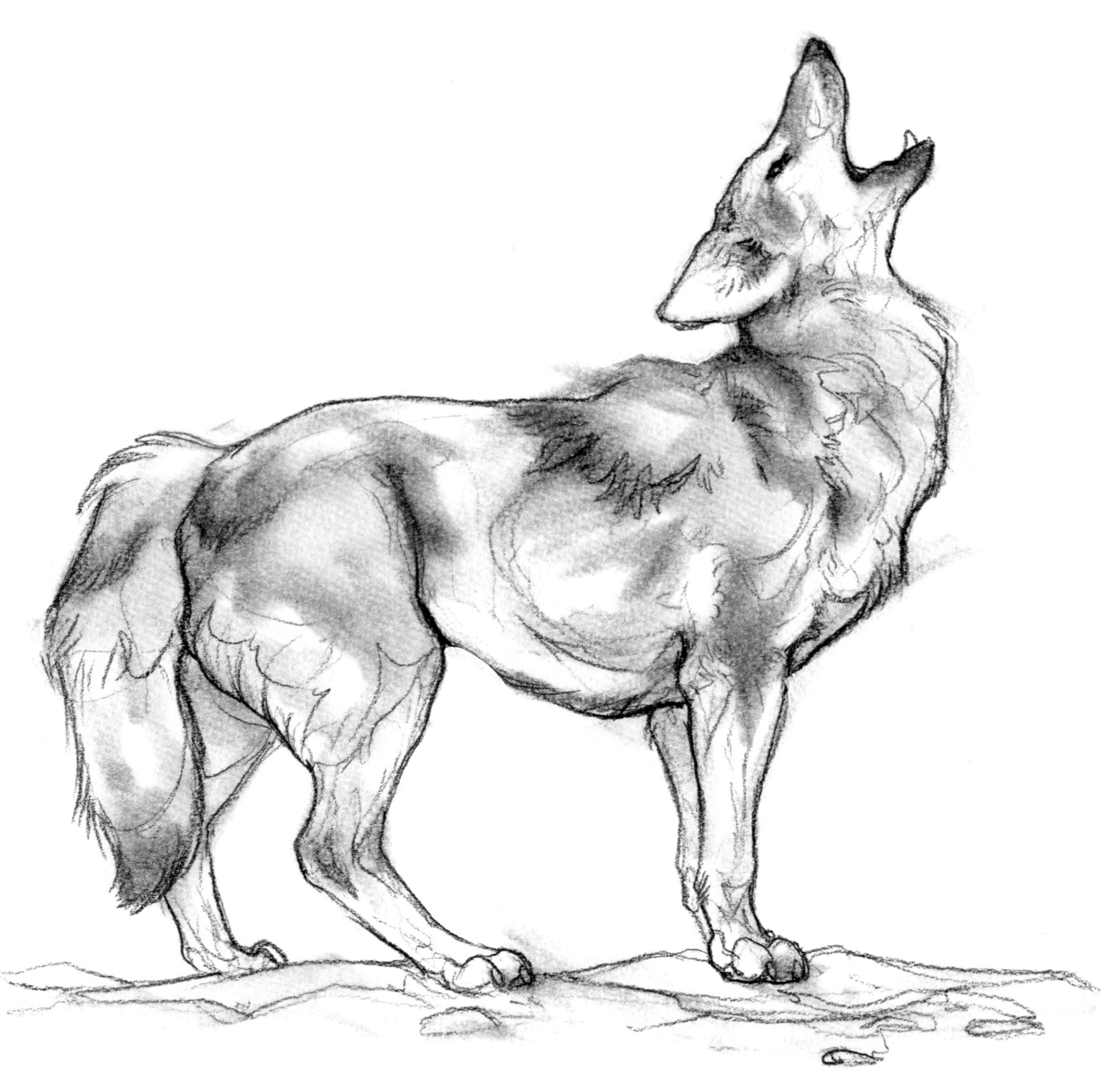

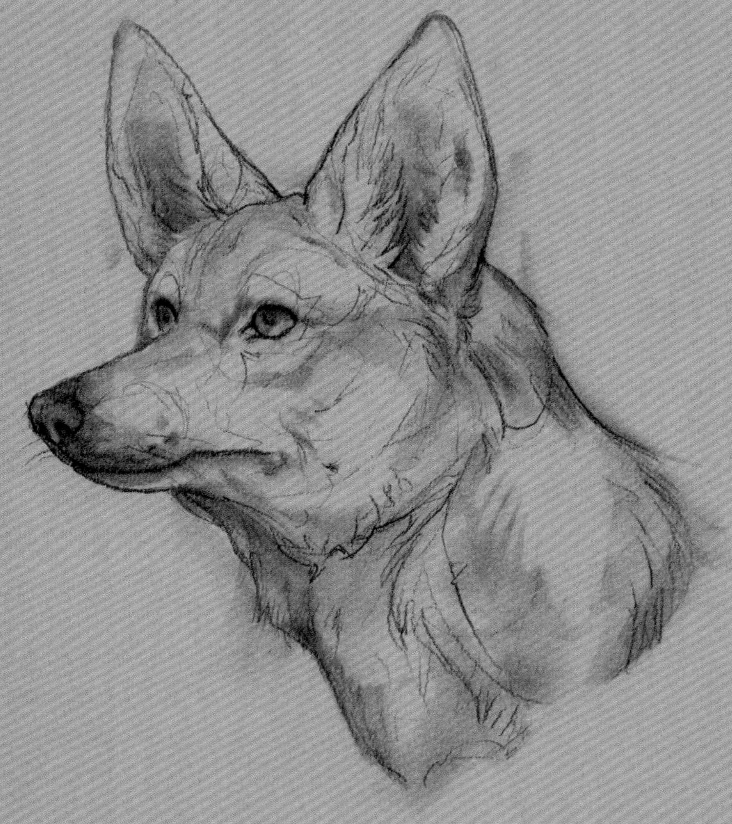

COYOTES

Location: Anywhere and everywhere.

Considerations: Keep your wits about you and leave the small dog at home.

Probably the most iconic animal of the desert—maybe after the rattlesnake. These resourceful creatures have called many lands home. They thrive all over North America and turn up in stories from many cultures. They're known for their cunning, their humor, and their elusive beauty.

Most folks who live in the desert know coyotes as omnipresent creatures, and they can mean a variety of things. For some, they are a noble friend who shows up along the edges of a lonely road—a smart and clever animal that has carved out a niche for itself among the thorns and rifles of its existence. For others, it is a thief, a killer of livestock, a pest that should be killed on sight. Either way, whichever side you fall on, the coyote is a common character in the North American desert.

If you've ever heard coyotes howling in the night, you know that it can be a chilling sound. Listening in on their conversation will usually leave you and your dog, sitting in silence, exchanging worried glances.

Coyotes are typically nocturnal around human development. Every so often, you'll spot one trotting along a highway or darting behind a dumpster beside an empty strip mall. Either way, seeing one during the day is like seeing a ghost. Your breath is shortened.

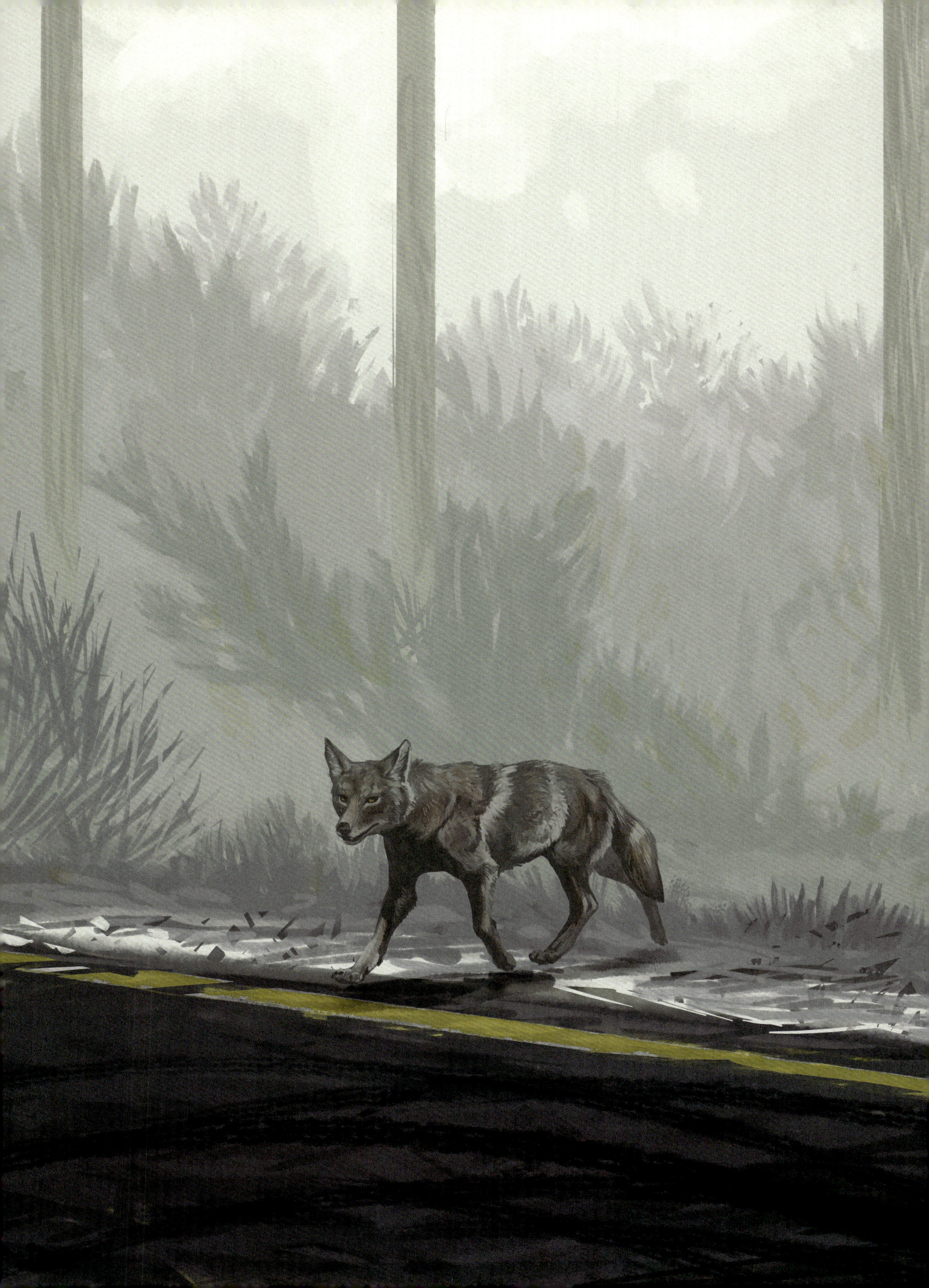

THE COYOTE MAKES EYE CONTACT. IT'S NOT LIKE MOST ANIMALS, WHO MAKE A BIG FUSS ABOUT IT WHEN LOOKED AT. THE COYOTE GLANCES AT YOU AND MOVES ON, SEEMINGLY TIPPING HER HAT. AS YOU DRIVE PAST, SHE KNOWS WHICH SIDE THE DRIVER IS ON.

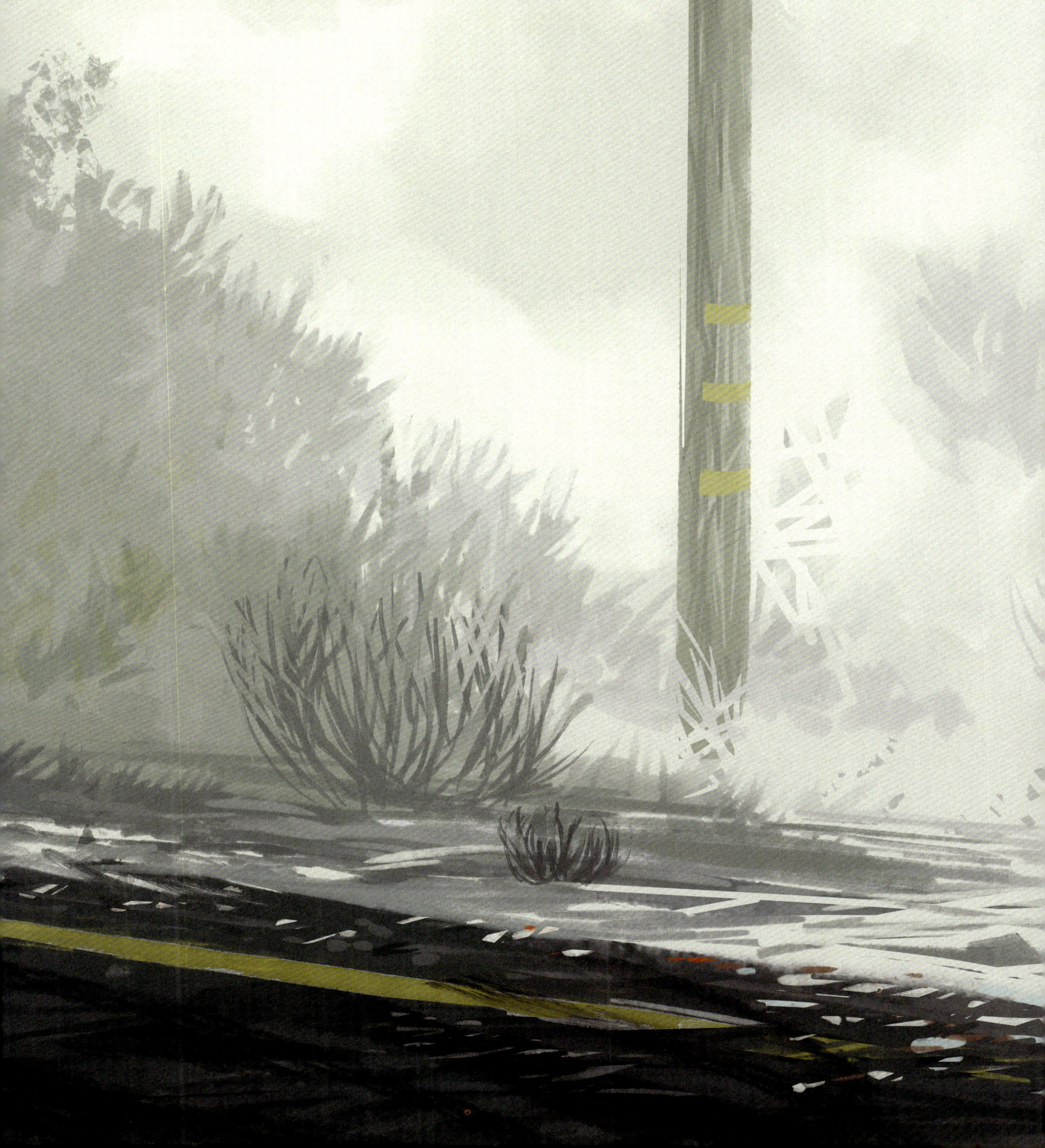

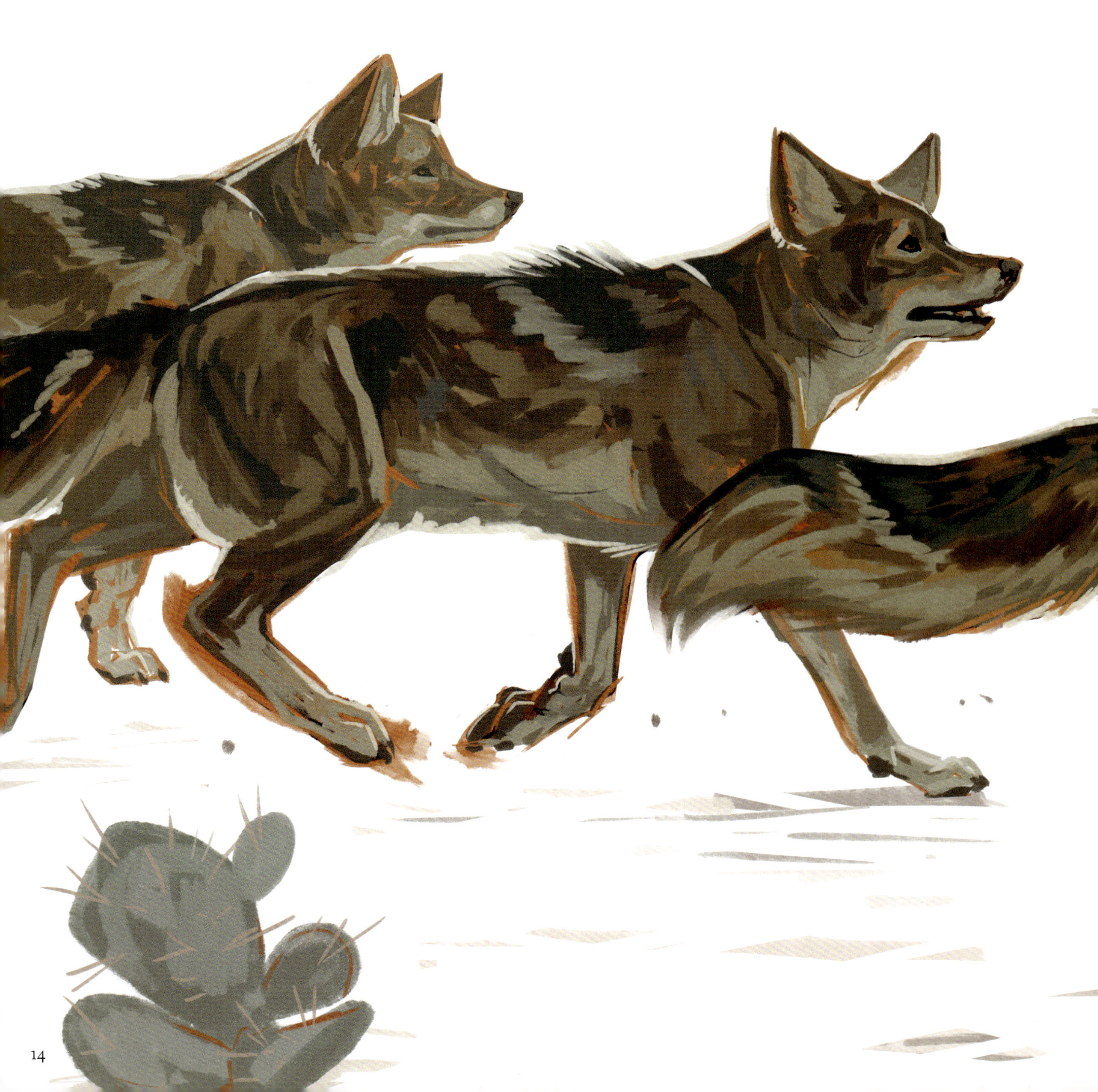

MANY WOULD AKIN THE COYOTE TO AN OUTLAW. UNTIL THE MIDDLE OF THE 20TH CENTURY, MANY STATES PAID BOUNTIES FOR COYOTES.

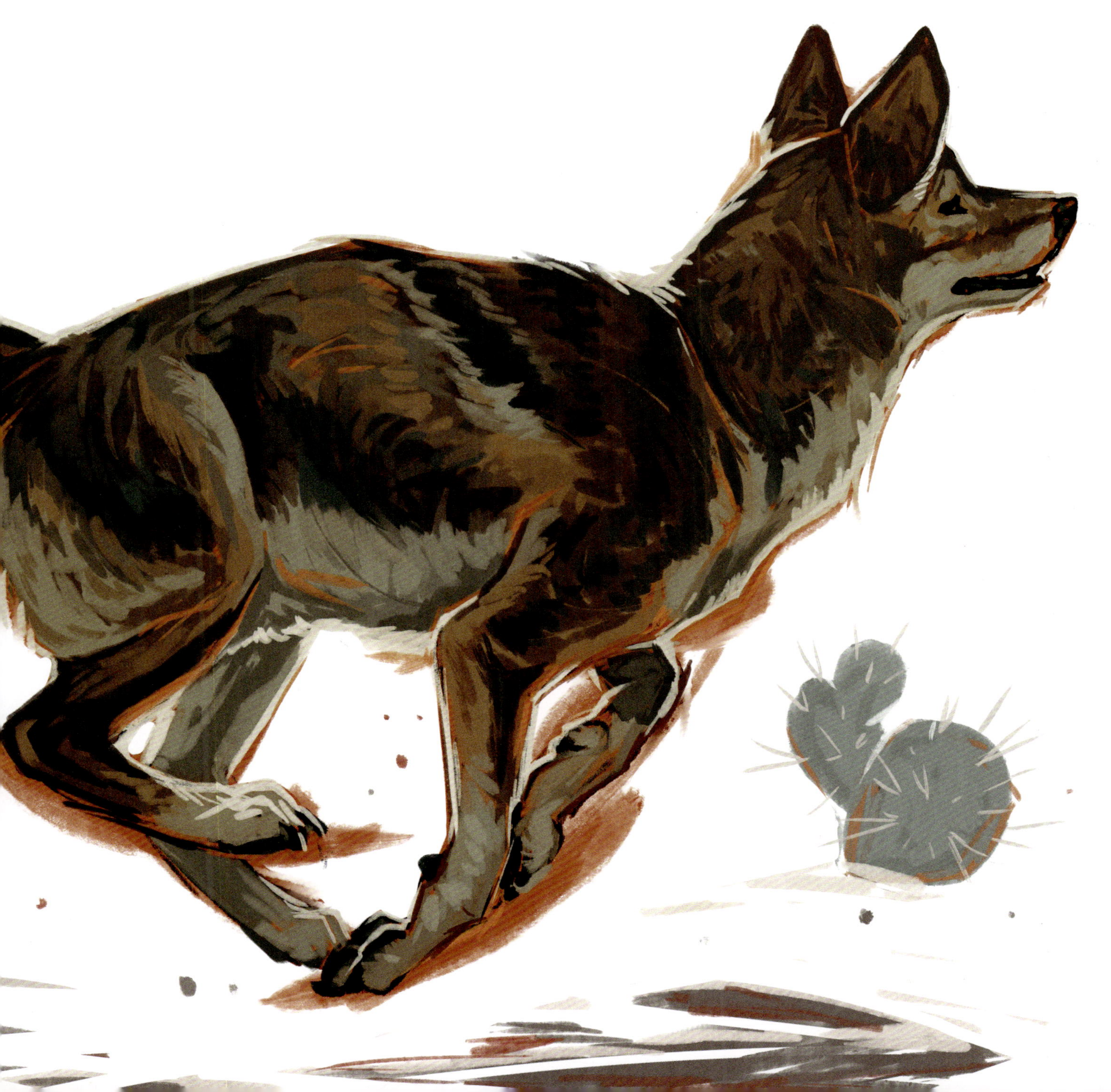

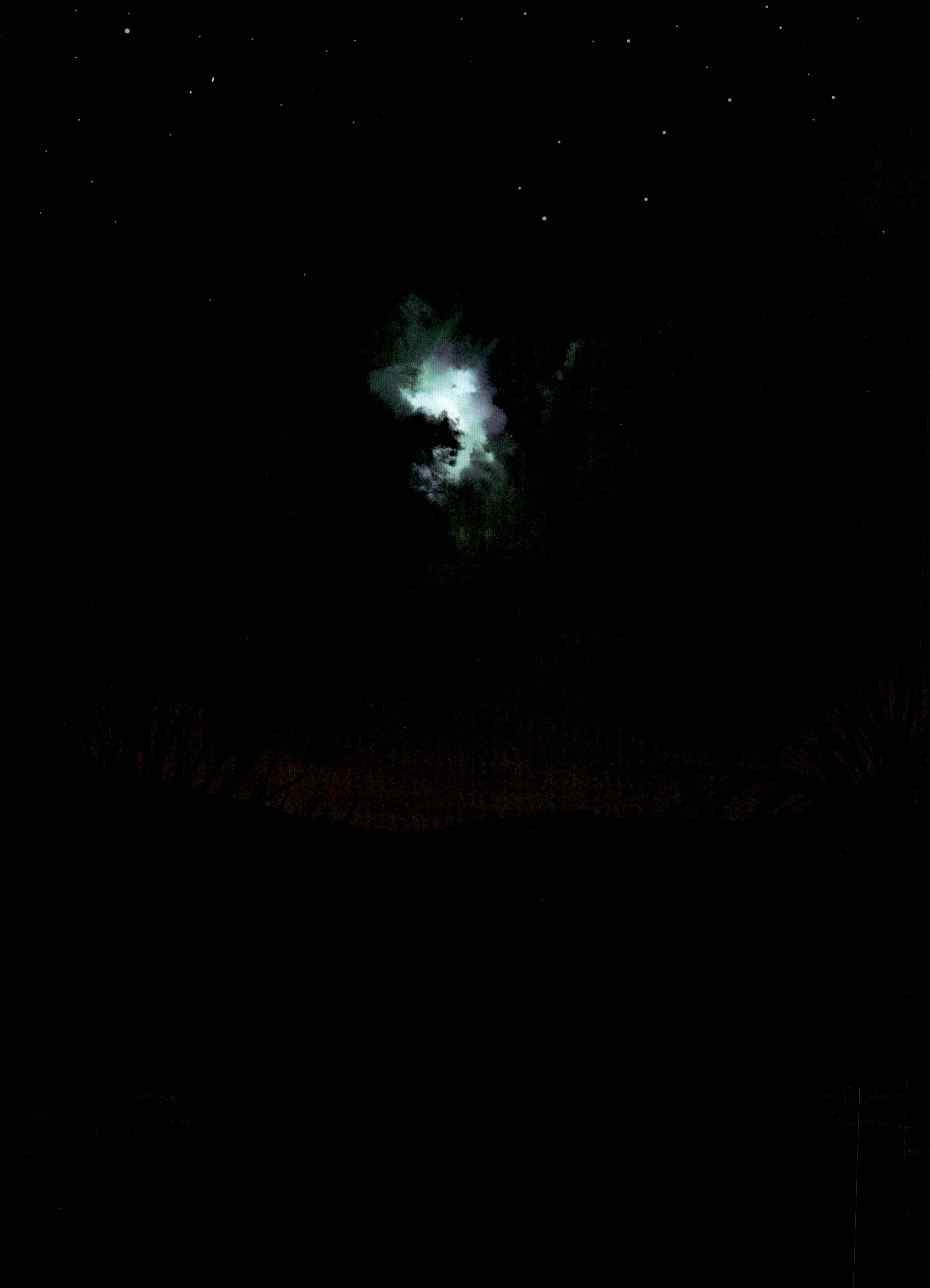

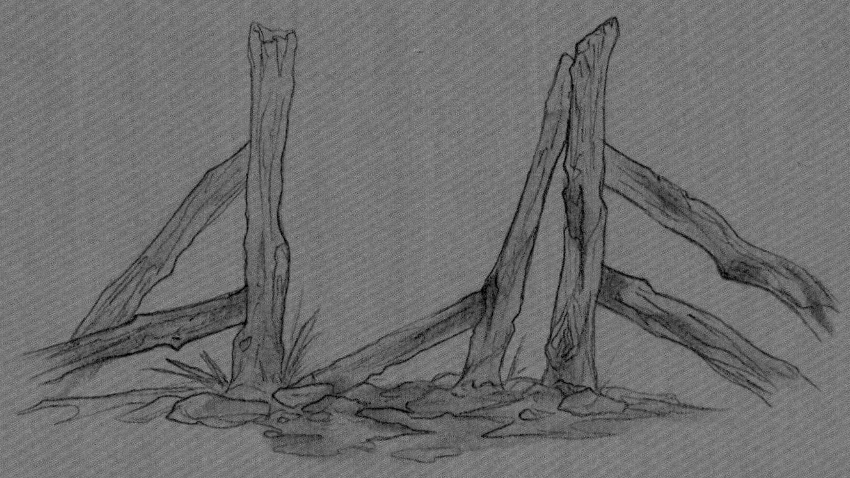

LIGHTS

Location: Mojave Desert, 34.921944°N 117.8275°W.

Considerations: Edwards Air Force Base is nearby, so be mindful of where you're driving. They occupy the area. The lakebed can be hot during the summer months, so pack water and tell someone where you're going.

Everyone in the desert has a story about lights. It's not uncommon to hear somebody at the local bar go on about how they encountered a ball of blazing white fire on the highway or out behind their shed. The desert has always been visited by lights.

When my grandmother was young and growing up in the hills of Roswell, New Mexico, she was visited by a light. This would have been in the early 1940s. *War of the Worlds* had been broadcast to a live audience just two years before, and terrible tales of aliens and creatures from other worlds would have been lurking in everyone's subconscious.

The way she recalled that night to me was that, as she was out playing in the twilight hours of a warm summer night, she spotted a green light on the horizon, just sitting there—ablaze and brilliant. She stood there and watched it, wondering what it could be. Suddenly, as if the light had noticed her watching, it barreled toward her. She ran. The fiery light chased her over the hillside, staying on her heels, sizzling and popping behind her, pursuing her with intent and intelligence. As she came over the hill, she cut her foot on an exposed tree root and fell to the ground. The light flew over her back and, with a loud pop, ascended into the darkened sky while whisking dust up into the air. She lay there for a minute, watching to see if it would come back down to earth, but it had disappeared. It was quiet. She got up, dusted herself off and walked home with a blood-soaked foot, a reminder of the strange encounter. The lights never went away, but she wasn't chased again.

The Mojave had its own lights. On summer nights, we'd spot what looked like headlights out in the dark. If you go out to that dry lakebed and watch, they'll show up. They move in pairs as they tear through the landscape. They start on one end of the lakebed and race to the other end, flying through the dust. They blink out of sight when they reach the end and will sometimes show up again for another round or two. So far we've counted upwards of eight individuals out there. Sometimes they show up without lights, and all you can see is dust reflected in the moonlight. They're silent as far as we can tell, but we don't go near them. Every now and again, we can hear a faint radio crackle out among them, like tuning a radio to find a frequency. "Racers" we call them.

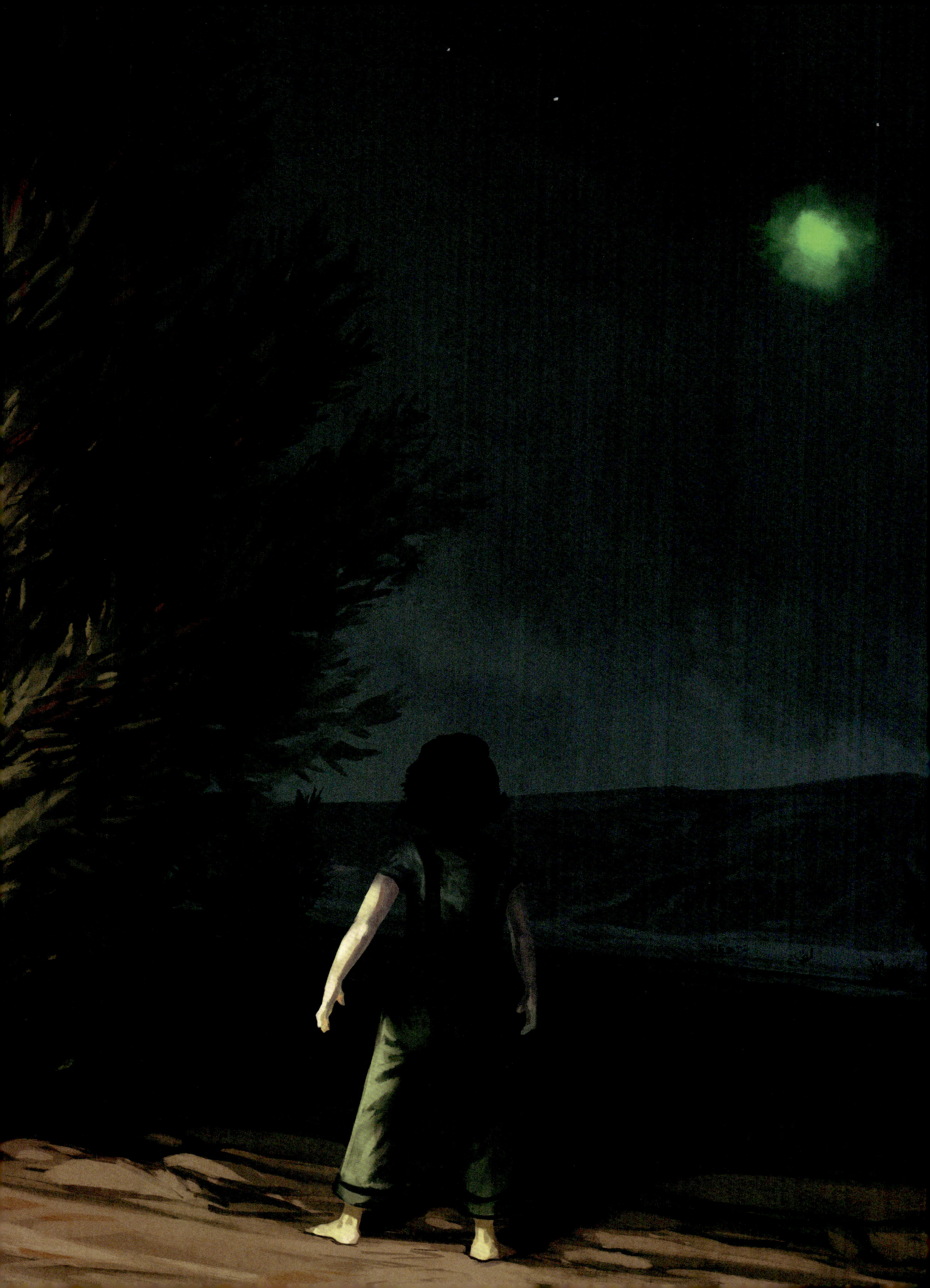

WHEN MY GRANDMOTHER WAS YOUNG AND GROWING UP IN THE HILLS OF ROSWELL, NEW MEXICO, SHE WAS VISITED BY A LIGHT.

DRY LAKEBEDS IN SOUTHERN CALIFORNIA HAVE BEEN USED FOR RACING BUT ALSO FOR EXPERIMENTAL AIRCRAFT. IN 1947, A MAN, PILOTING A ROCKET ENGINE POWERED AIRCRAFT, BROKE THE SOUND BARRIER.

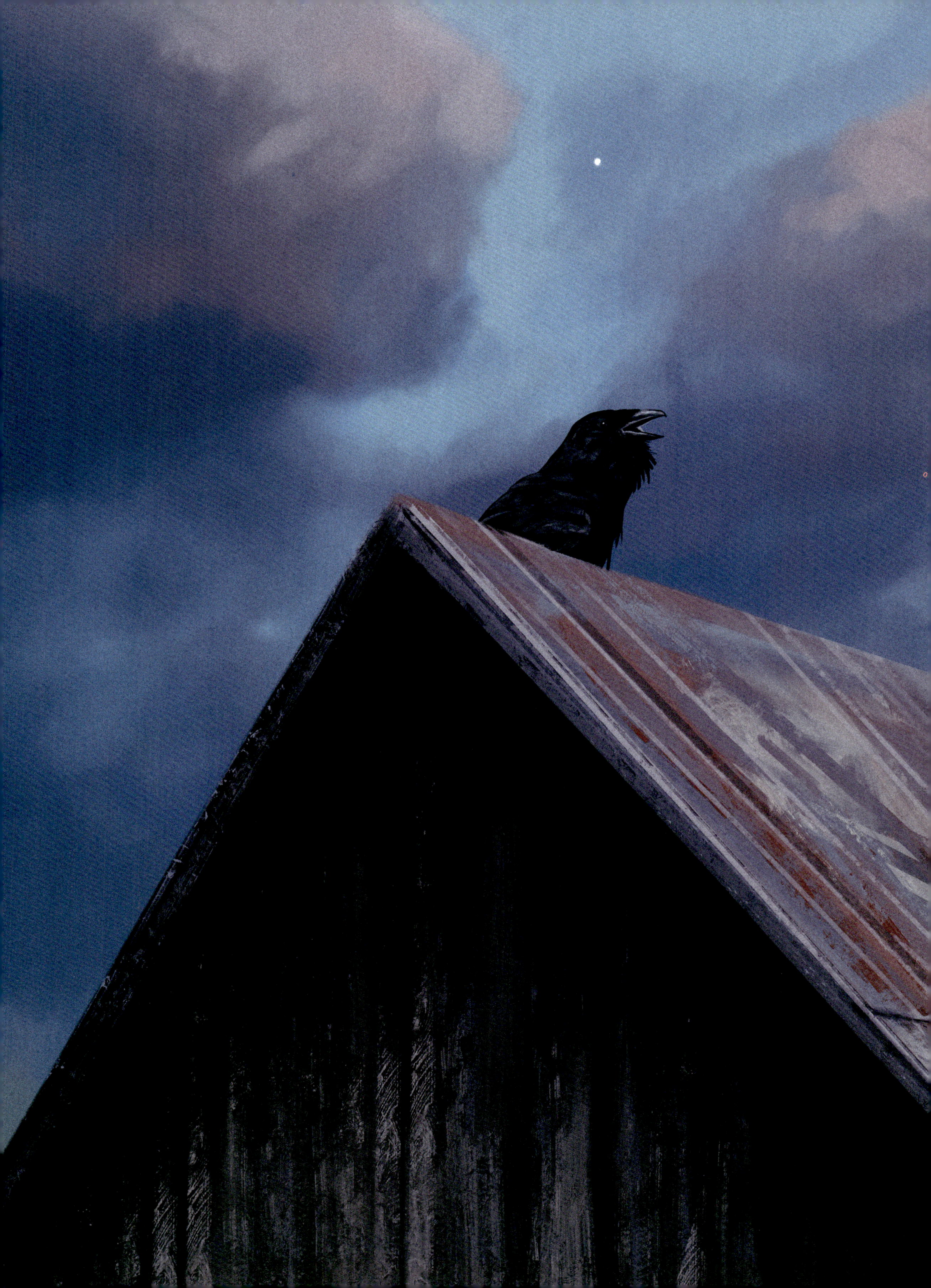

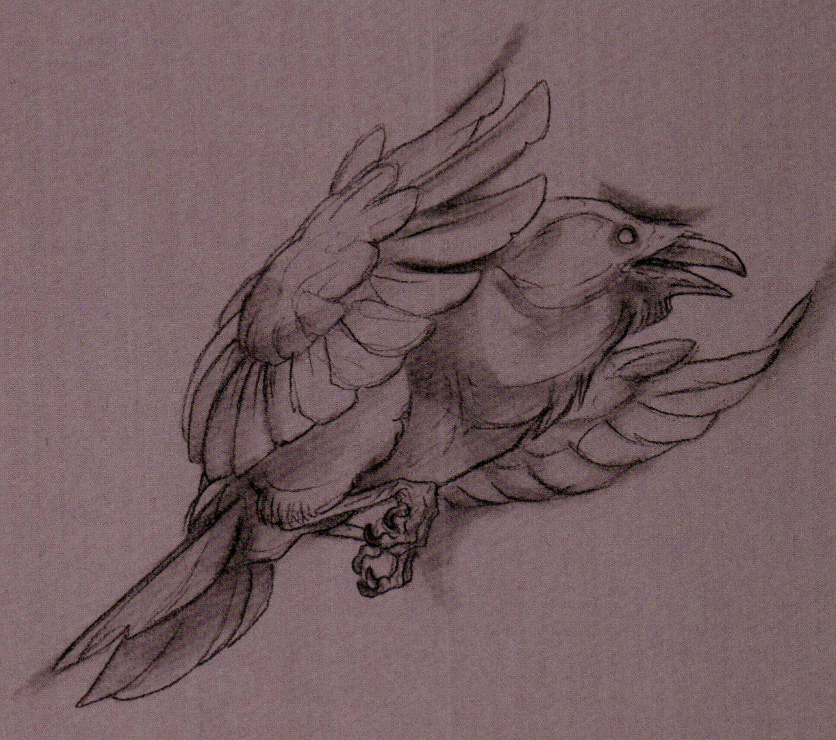

RAVENS

Location: All throughout the desert southwest and beyond.

Considerations: Ravens are very intelligent birds and can remember faces. Be on your best behavior.

Ravens have a range that spans many continents, and they are native to the Mojave Desert. They are some of the most intelligent birds in the world. They belong to the family known as Corvids and are known for their playful nature and quick wit. They're one of the only birds that can make tools and remember faces. Because of their charismatic nature, they are of great importance among the Indigenous people of North America and other cultures all over the world.

The following is a story recalled to me from a small homestead in Nevada:

The ravens of Venus Ranch are an interesting bunch. It seems we have a few individuals who return every couple of weeks to check up on the place. At least we think so. In any case, there are a few that hang near the goat barn, and they have the same mannerisms and habits as before. Could be learned. They all make their usual raven noises and do their usual raven things, but one of them makes a particular croaking noise with its throat that stands out among the rest.

It's not like the usual high-pitched "cockle clack" that ravens make. It's a dull and deep croak that emerges as though the bird has some sort of tar in its throat. This individual typically makes the noise at dusk, when the stars just start to peek out of the sky and the last wisps of hot air begin to recede with the sun. We've nicknamed this raven Dylan.

Other raven sightings usually depend on what dead thing has shown up on the property. Every so often an animal dies, and the ravens are the first to let us know. One early morning, we were awakened by their calls. Outside we could see them, clamoring for one of the goats. Animals die on ranches all the time. It's not uncommon.

The morning was brisk. Autumn was just about to begin, and the nights were letting go of their heat. Rock and sand crunched beneath my feet as I walked to the barn to grab my gloves. I could see Dylan sitting on the peak of the barn roof. He seemed to be sitting this one out.

The other ravens were squabbling and cawing at each other in the yard around the carcass, pieces of gravel and fur flying about. A goat lay on its side, burst open from belly to chest. Strange. I could make out something shiny inside the carcass—small, irregular black fragments. There were hundreds of them stuffed inside the cavity of the body and spilling out onto the dirt. The ravens were plucking them up frantically and swallowing them, one after another after another.

I was able to snatch one from the tangle of black wings and beaks. The surface of the object was smooth, with small bits of rust and small irregular holes dotting the surface. It was heavy for its size. A meteorite.

What were hundreds of them doing inside of one of the goats?

I walked back to the barn to prepare for disposal. Dylan was still sitting up on the peak of the barn roof. He knocked at me once or twice and sat there, peering.

I stood there, looking at him, his feathers shiny and black against the periwinkle sky of the morning. I held the meteorite in my hand and wondered if maybe he'd had his fill.

—Eliza James, Venus Ranch

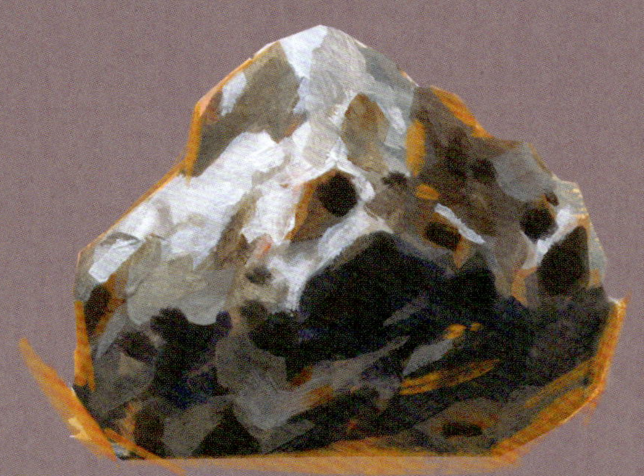

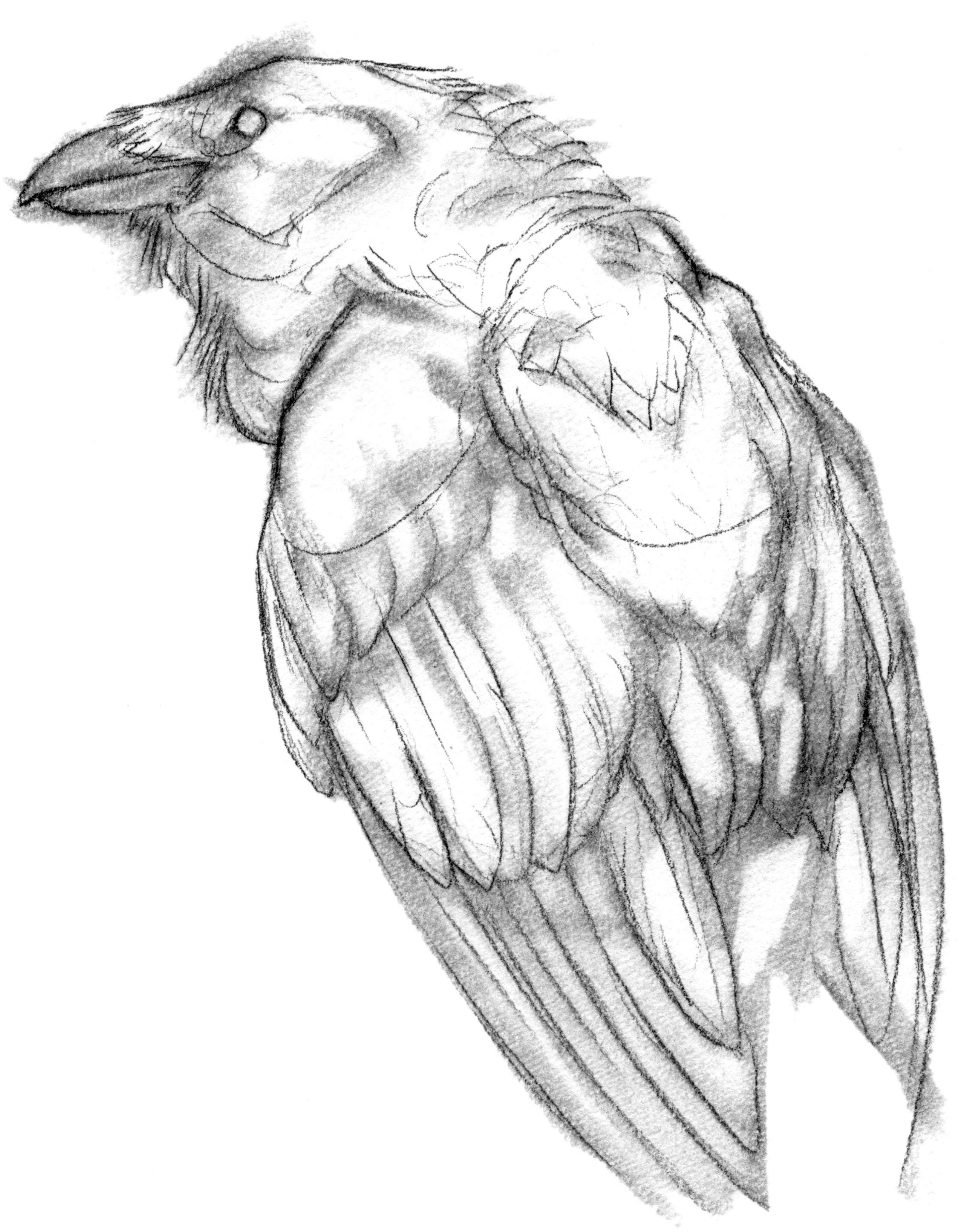

RAVENS CAN IMITATE HUMAN SPEECH AND MANY OTHER SPECIES.

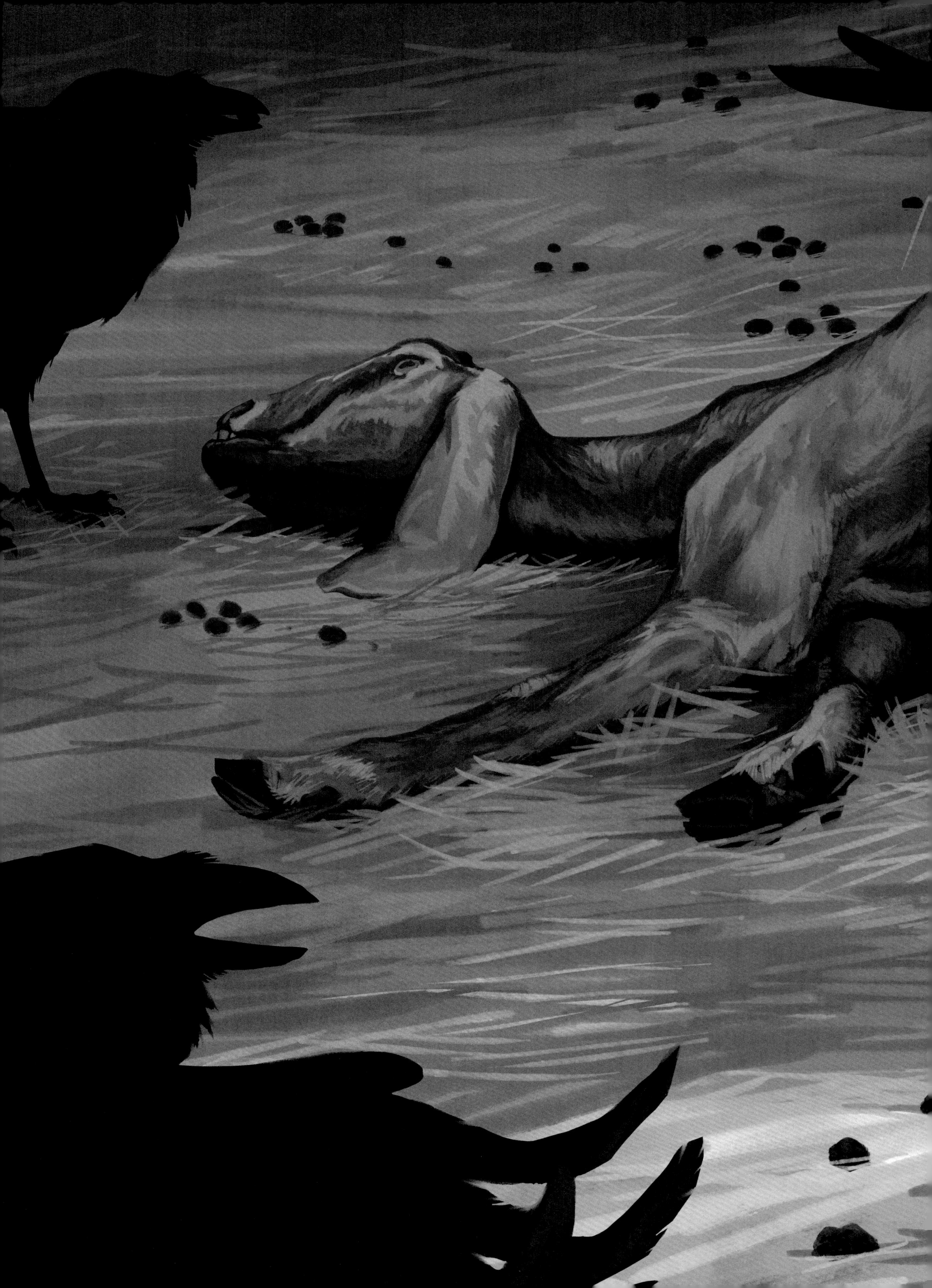

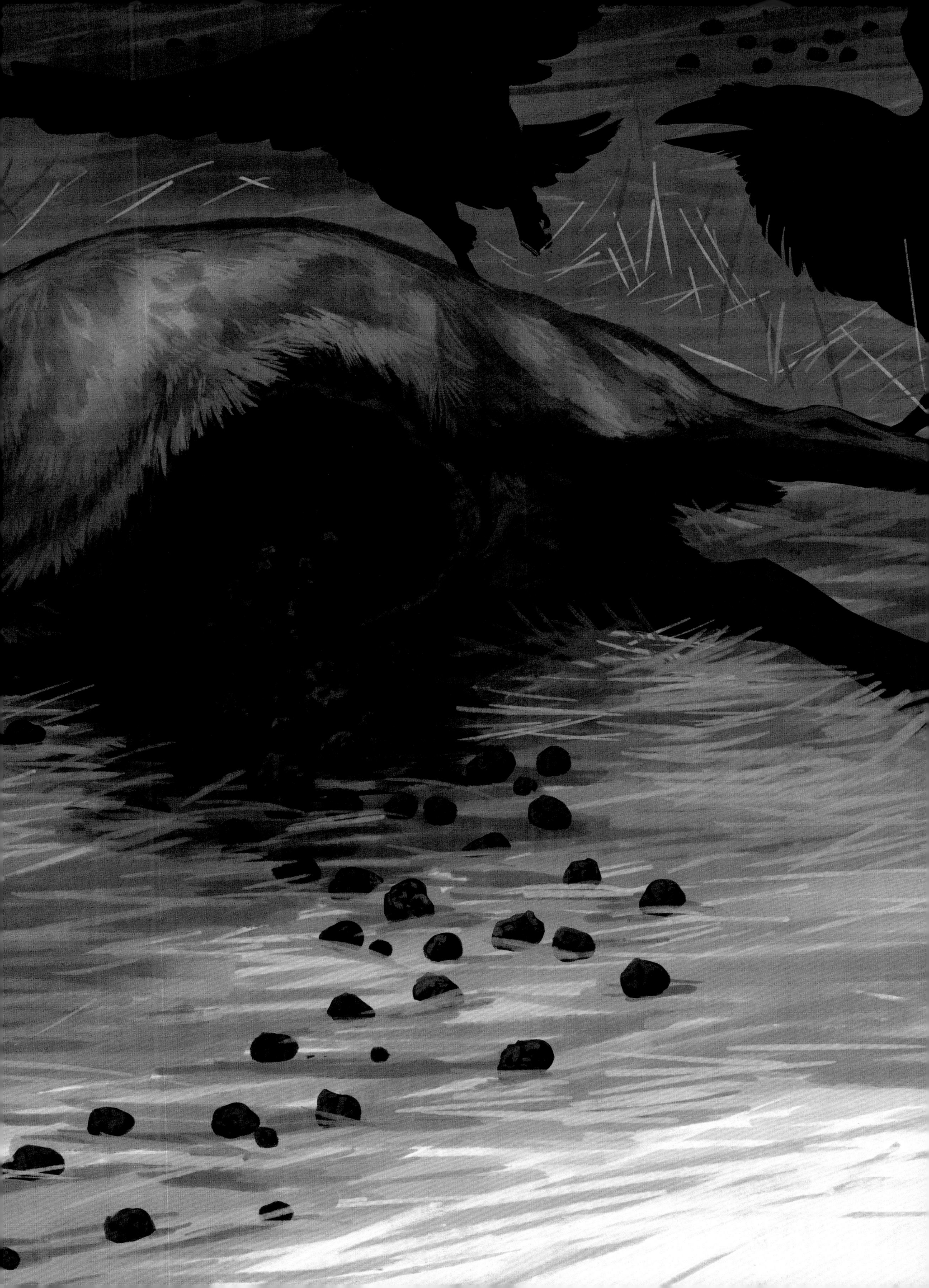

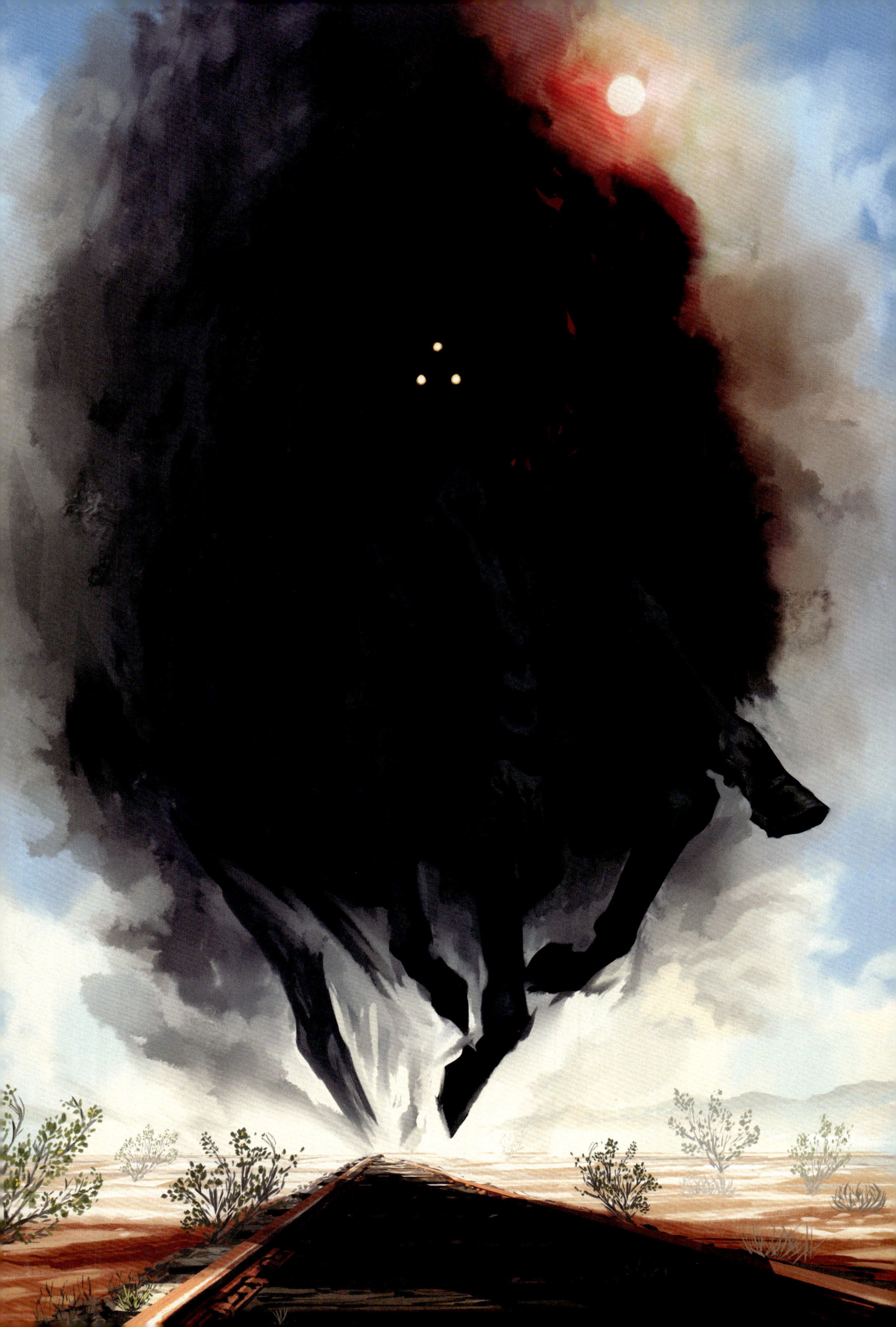

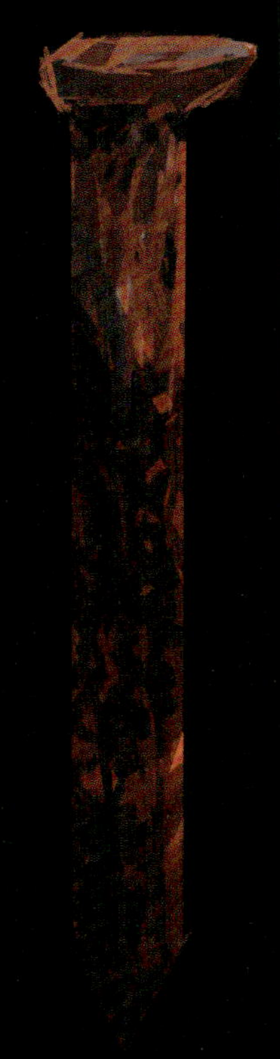

RAILROADS

Location: Southern Arizona railroads.

Considerations: This encounter is best left alone; the sound can be deafening.

Along those dusty red railways in southern Arizona, a specter of despair and anger barrels across the desert landscape. Tangled limbs and eyes ablaze, it charges relentlessly along the metal and wooden scars that outline the Southwest. As it tears across the tracks, a booming roar can be heard, like thunder colliding with metal.

Many ghosts walk along the old railways of the West. Many were forced to perish in their creation.

Those who have seen it left the encounter with little or no hearing left. It's best not to wander those lonely rails. It may result in seeing the dreaded thing.

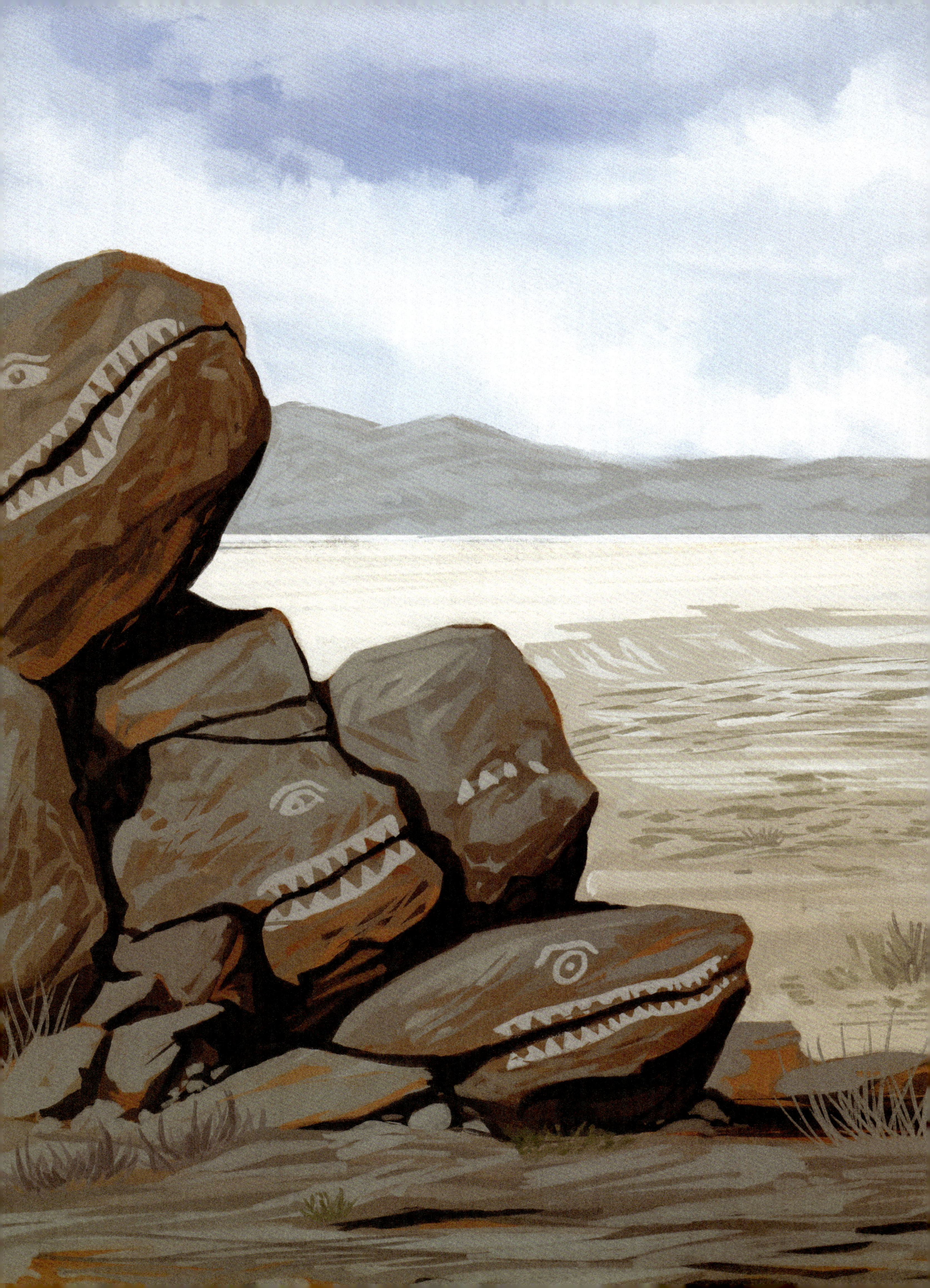

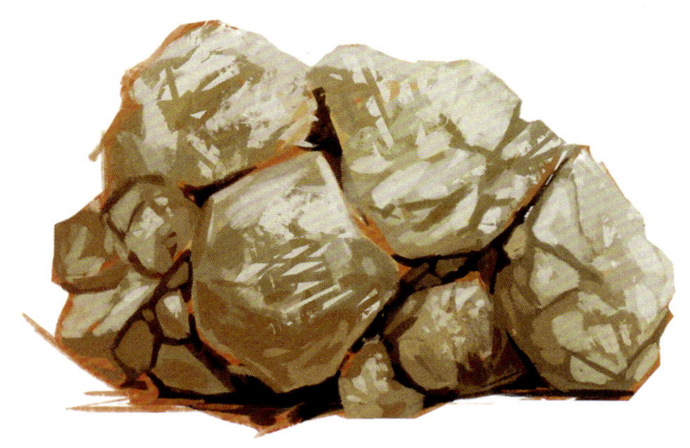

THE TRONA MULE

Location: The Trona Salt Flats.

Considerations: This specter is a rare occurrence. You'll have a harder time looking like you're up to nothing in front of the locals. It's best to keep moving.

Out on the scarred salt flats of Trona walks a mineral-encrusted mule. The specter can be spotted at sunrise, wandering about the steam vents and mineral planes of the dry lake. The mule showed up when prospectors were surveying the area for its worth.

It is thought to be a bad omen.

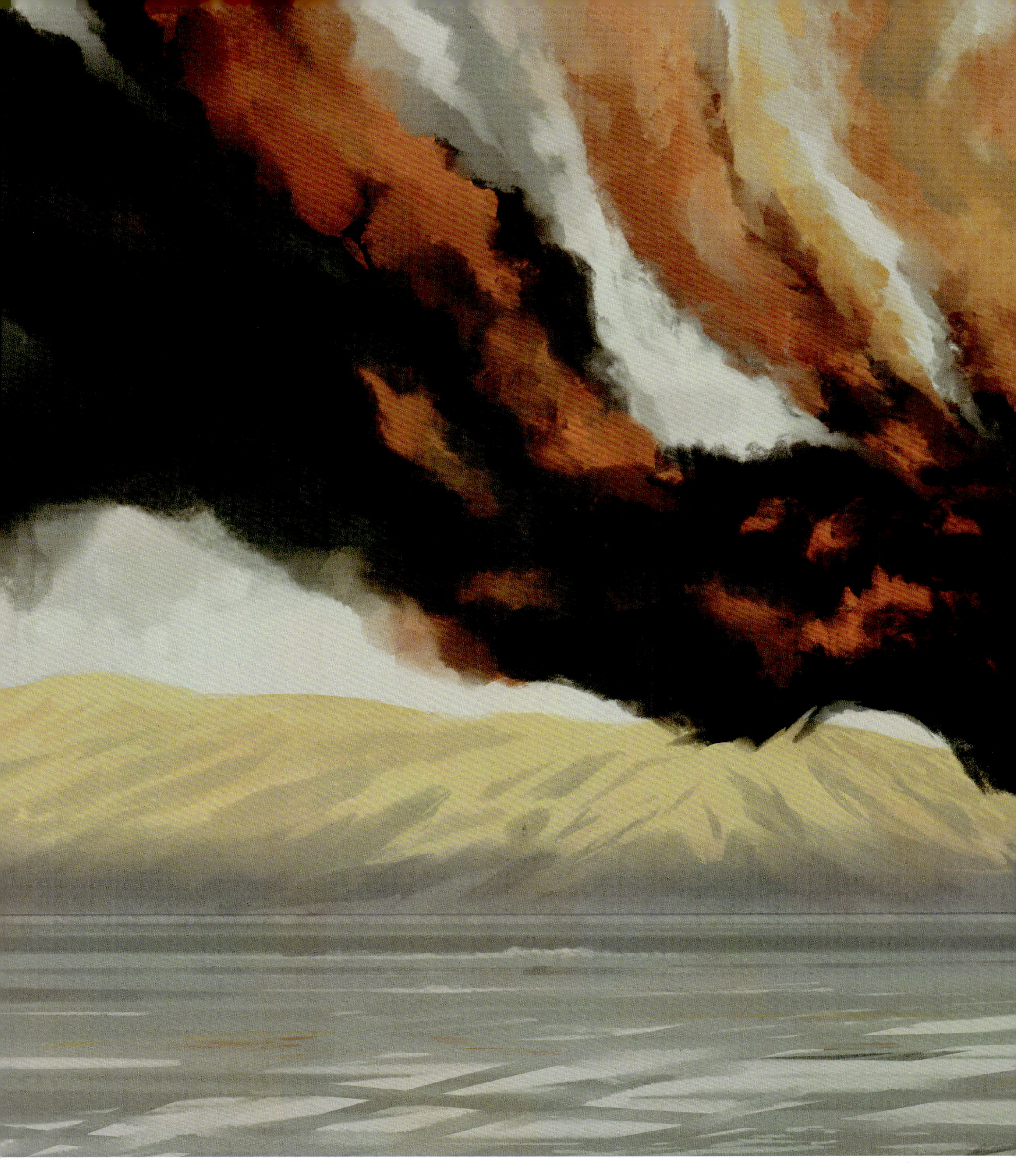

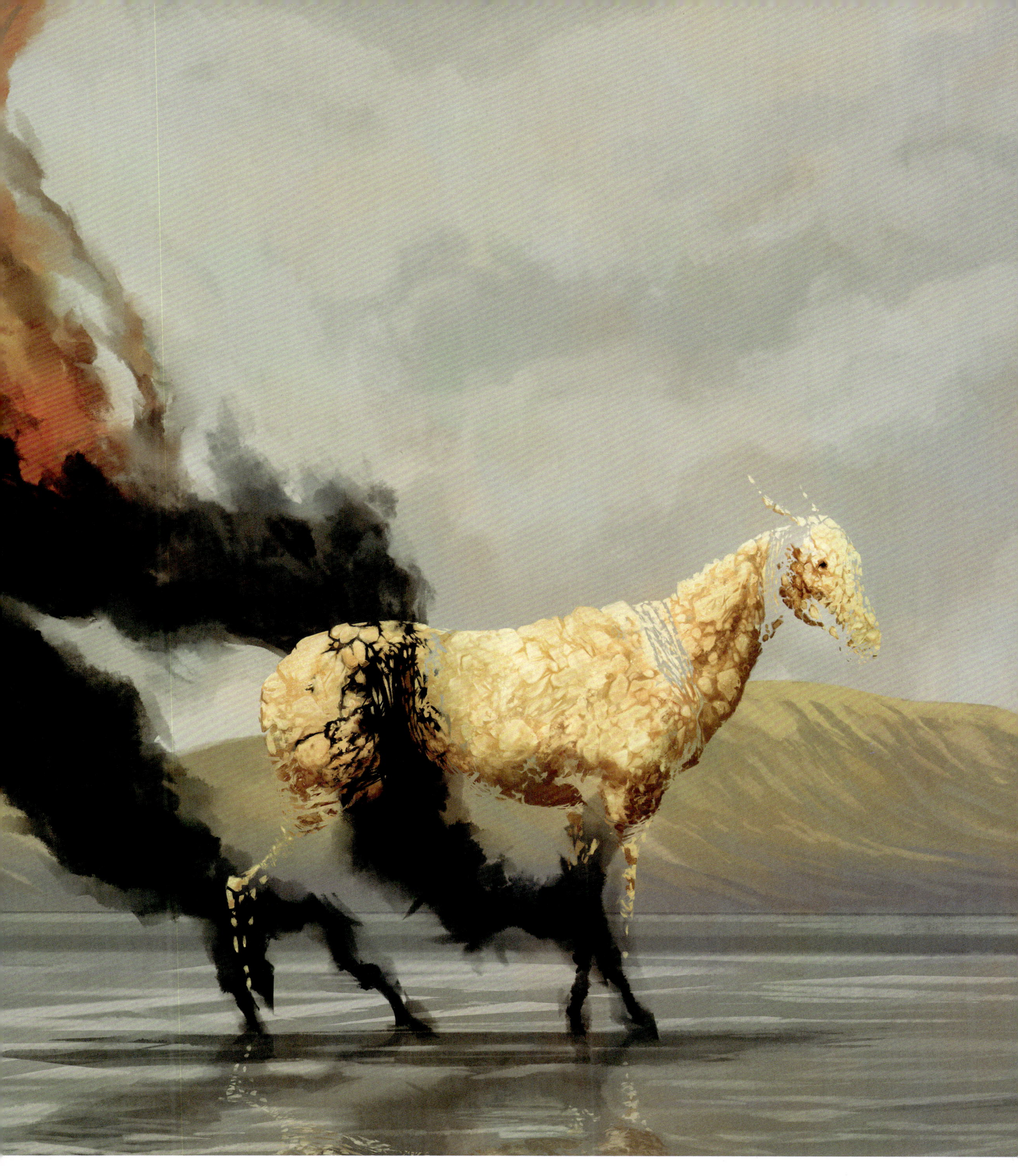

TRONA IS MINED AS THE PRIMARY SOURCE OF SODIUM BICARBONATE IN THE UNITED STATES.

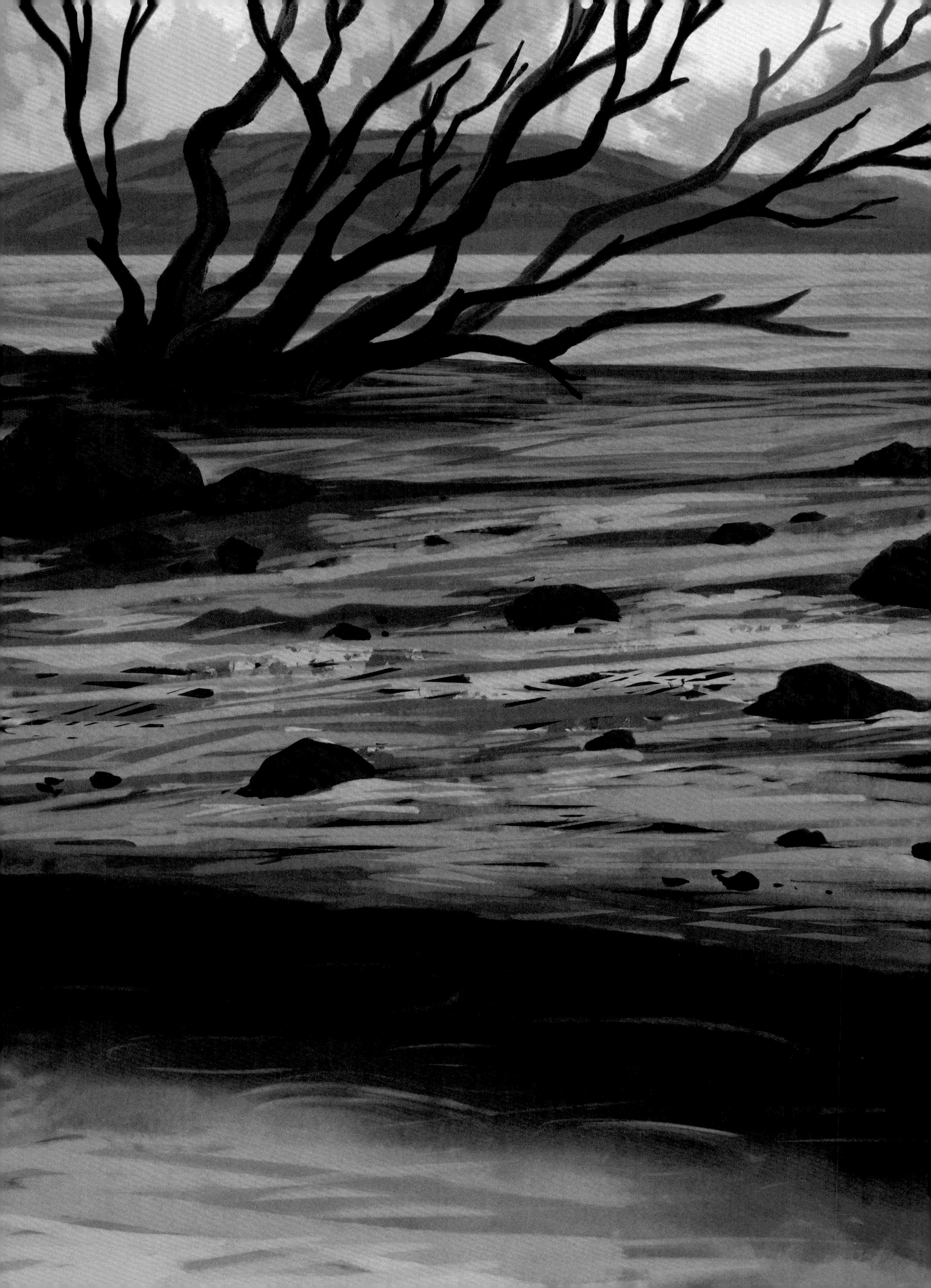

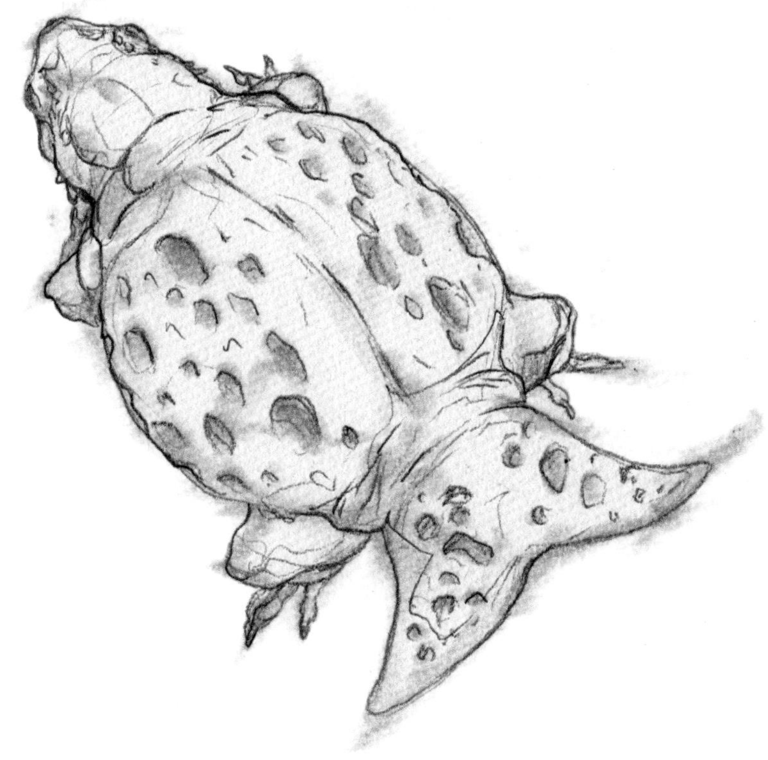

BLACK ROCK DRAGONS

Location: Old lava flows.

Considerations: These creatures are extremely rare and only appear before a rainstorm. Don't stick around too long. Floods can be a hazard.

In the old lava flows of the Mojave exist small creatures called Black Rock Dragons. They're not actual dragons, possessing no magic or breath of fire. However, their appearance is that of some ancient creature. With flat, hardened bodies, they scurry about the pitted lava rock strewn about the golden hills of the southeastern Mojave. Their skin appears to be pitted and shiny, to blend in with the rocks and lava beds. Some know them as Amboy Devils.

They pour out of the earth before the rain. As they lie there, they open their mouths, revealing a shiny black interior like the heart of the planet.

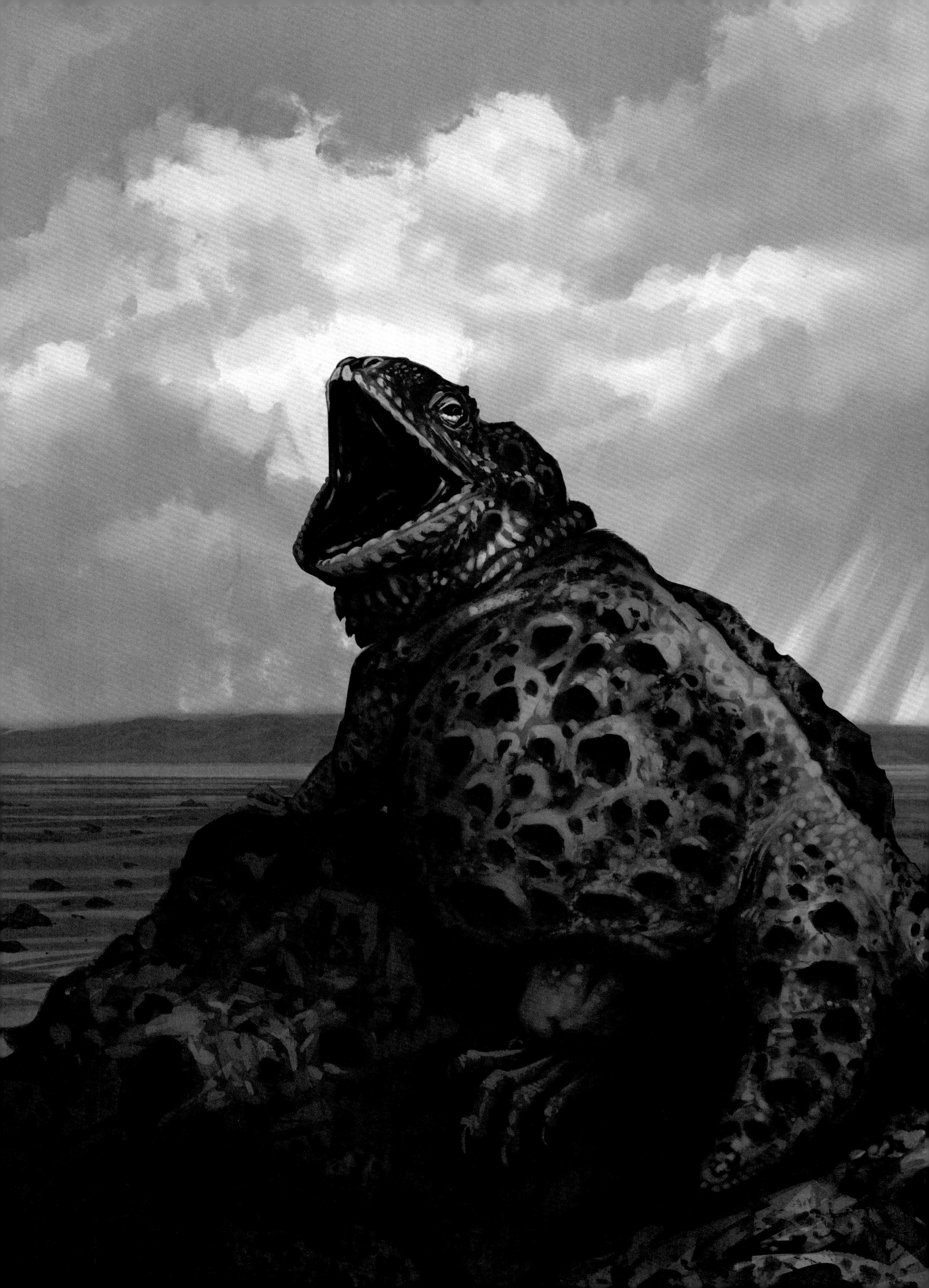

THESE CREATURES EXHIBIT A SPLIT
TAIL, MAKING THEM UNIQUE IN THE
REPTILE WORLD.

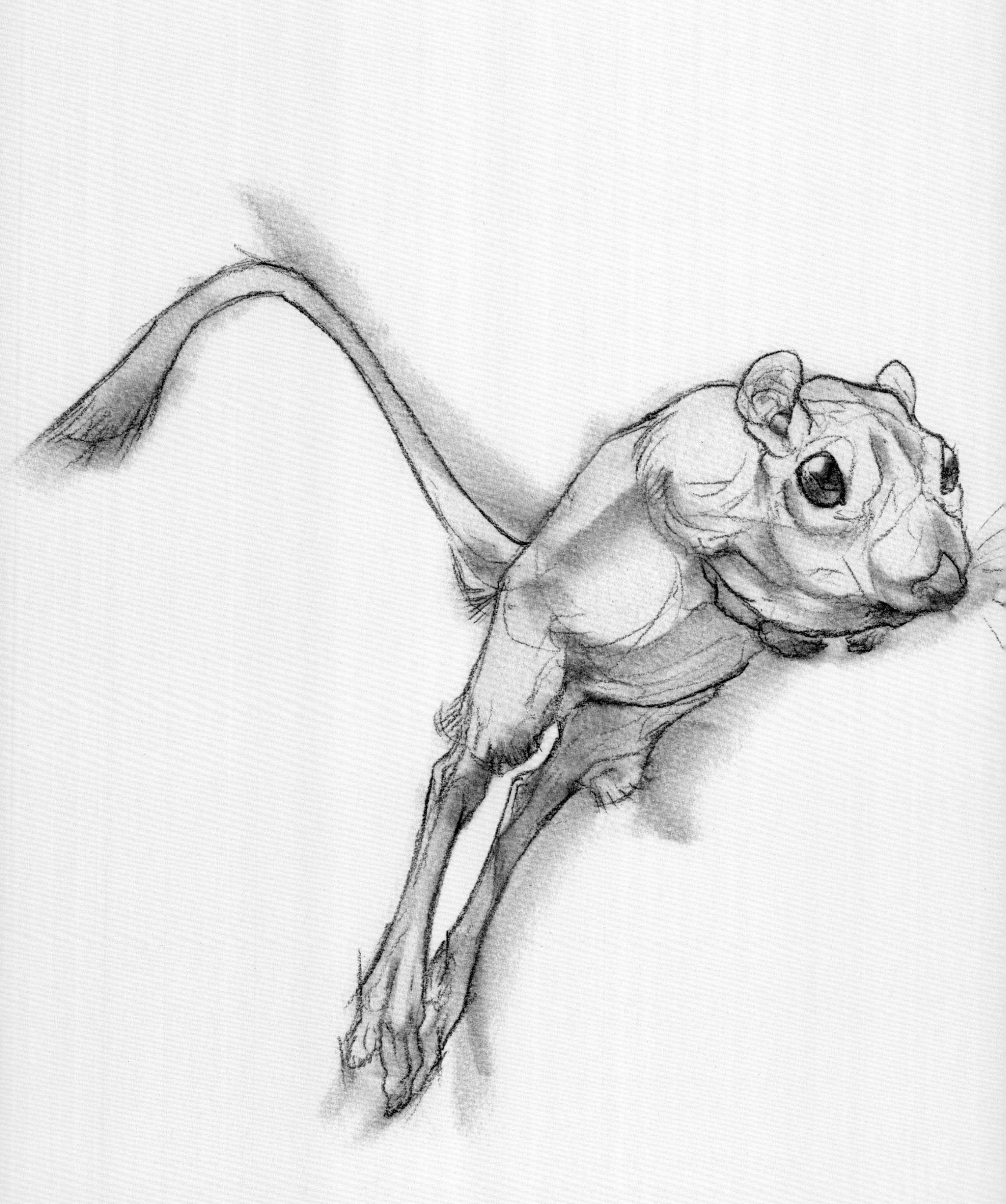

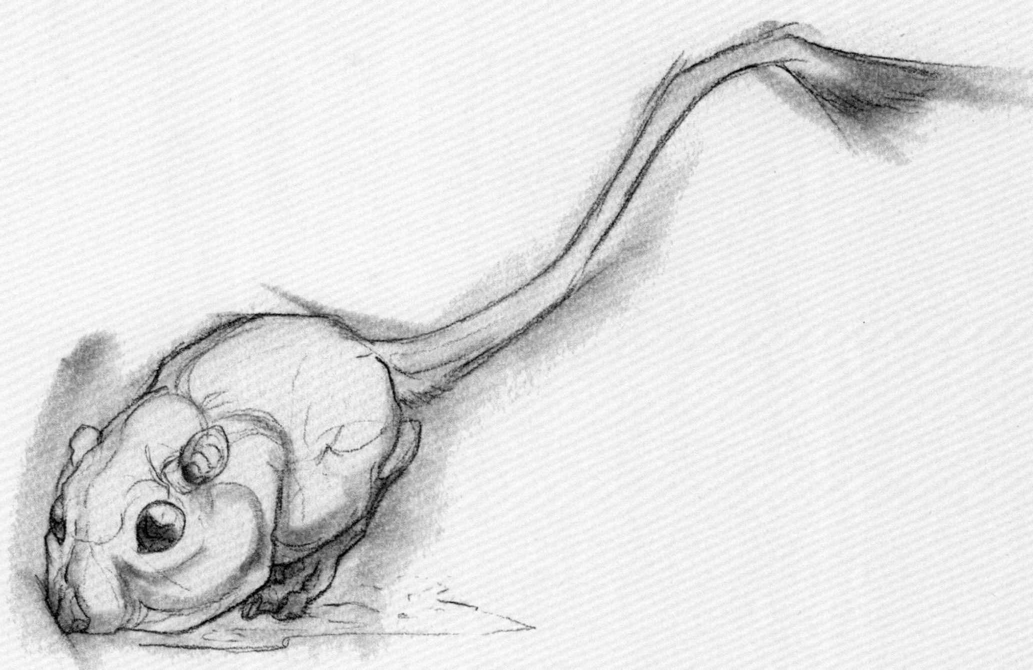

KANGAROO RATS

Location: Creosote bushes and other desert scrubs.

Considerations: Be mindful when looking for these rodents. Rattlesnakes and other dangerous animals could be provoked.

The desert kangaroo rat is a small rodent native to the Southwestern United States and Mexico. Its habitat can be found in the Sonoran Desert as well as the Mojave. It is characterized by its long, kangaroo-like hind legs and its peculiar hopping gait.

This little rodent is a survivalist master, living in some of the harshest places of the Mojave. They do not *need* to drink water, finding most of their requirements in the seeds they eat. Kangaroo rats can humidify their burrows using their body heat. When coupled with the cold temperatures of the soil, a side effect occurs: The stored seeds in their burrows absorb the ambient moisture, returning the water to the rodent.

Water is the most precious resource in the desert. Humans have carried, shipped, stolen and rerouted entire river systems to get water to the dry ground of the desert. So much about this animal sounds unbelievable, and yet these tiny rodents thrive without.

"I was reading by the campfire and happened to look down. The light of my headlamp caught two kangaroo rats. They were fighting—scrambling around and kicking each other. I wouldn't have noticed them otherwise. Their squabble was completely silent."

—Kate Marshall

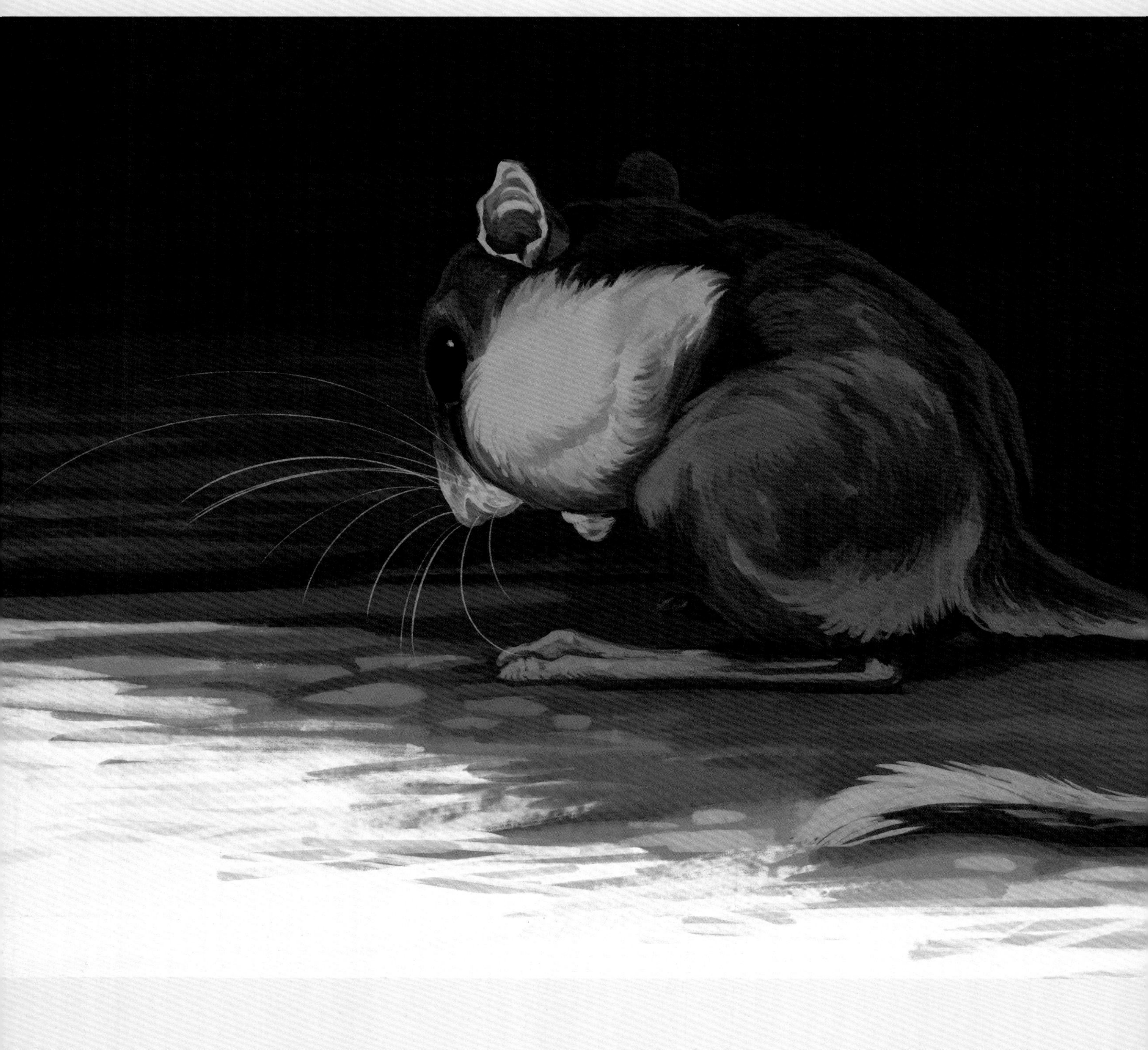

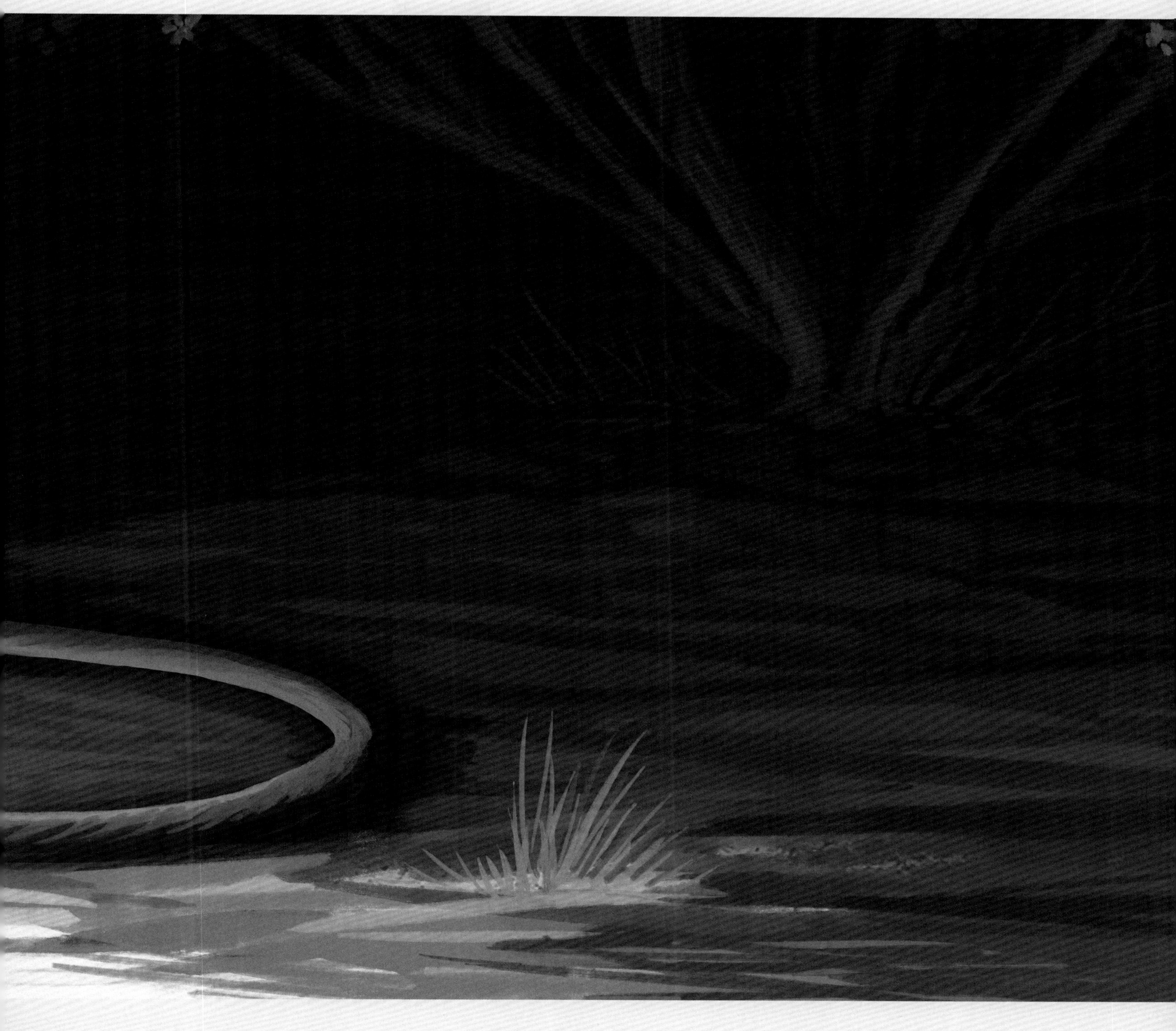

THE WARMTH OF A CAMPFIRE IN THE COLD OF WINTER WILL BRING OUT THESE LITTLE RODENTS.

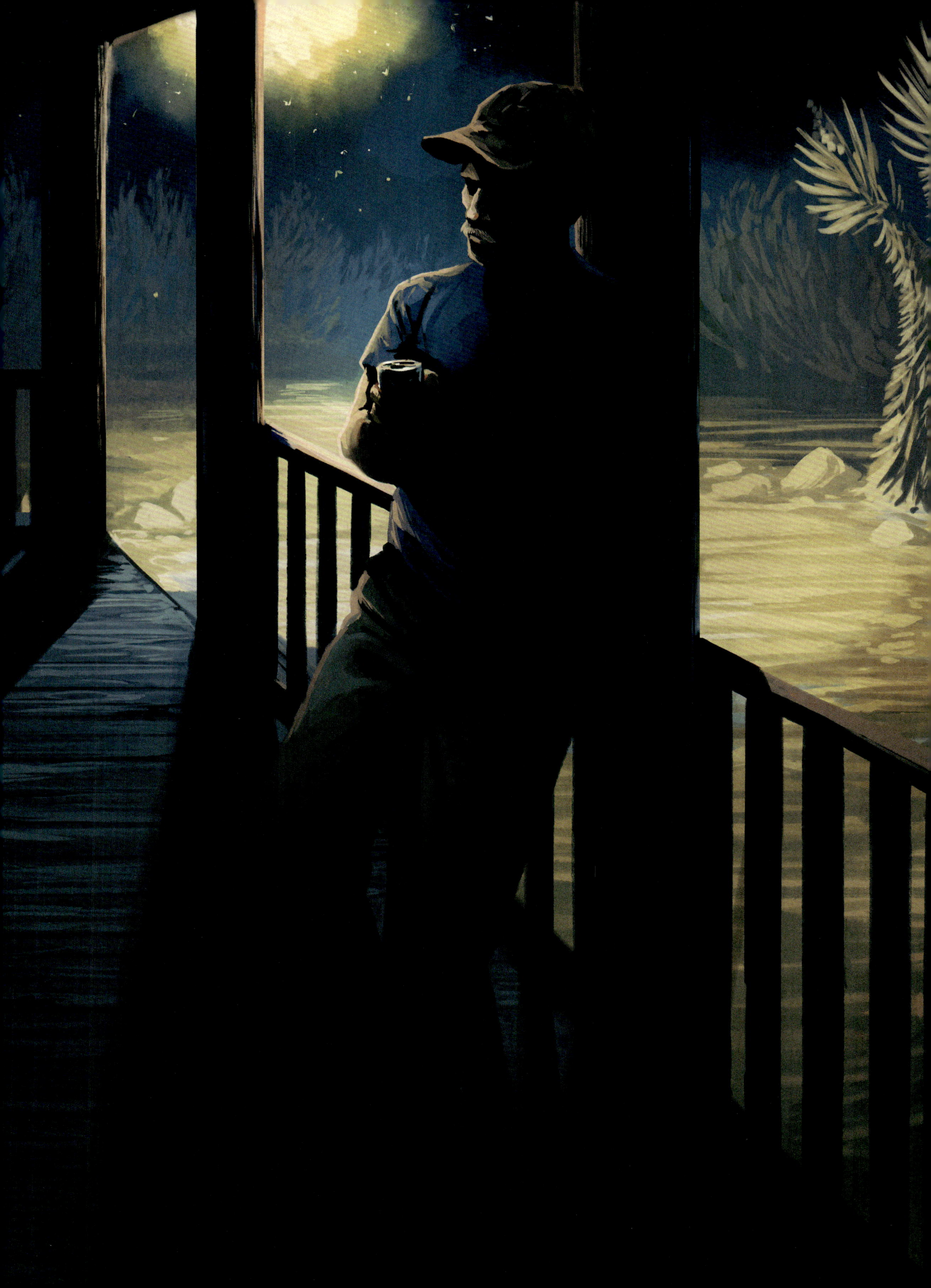

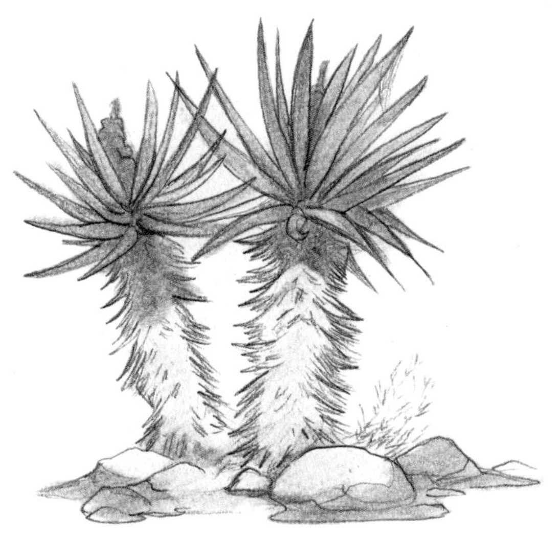

RED TOADS

Location: Omitted for the privacy of the inhabitants.

Considerations: This occurrence was told to me through correspondence. I took some liberties and changed some names.

It started on a Sunday evening. After we'd finished feeding the animals, we relaxed on the front porch taking a moment to enjoy the cool air that night. It had been hot lately, but there was rain coming.

We heard them, quiet at first: something stirring out there among the juniper trees toward the south end of the property. It was sort of a low-pitched sizzle, almost like the sound of bacon fat in a skillet. It was distant. At first, of course, we thought "FIRE," but there was no glow and no smell of smoke in the air. We sat there listening, hushing each other to try to hear more as we sipped our beer in stillness. It grew closer as we huddled our heads down, as if the posture would make our hearing better. Every three minutes or so, the sound would swell and then recede.

Amos got up to grab another beer from the cooler. As we drank, he stood against the railing and picked at the wooden post of the porch. We shared a nervous laugh every now and again. Just unsure. Spooked.

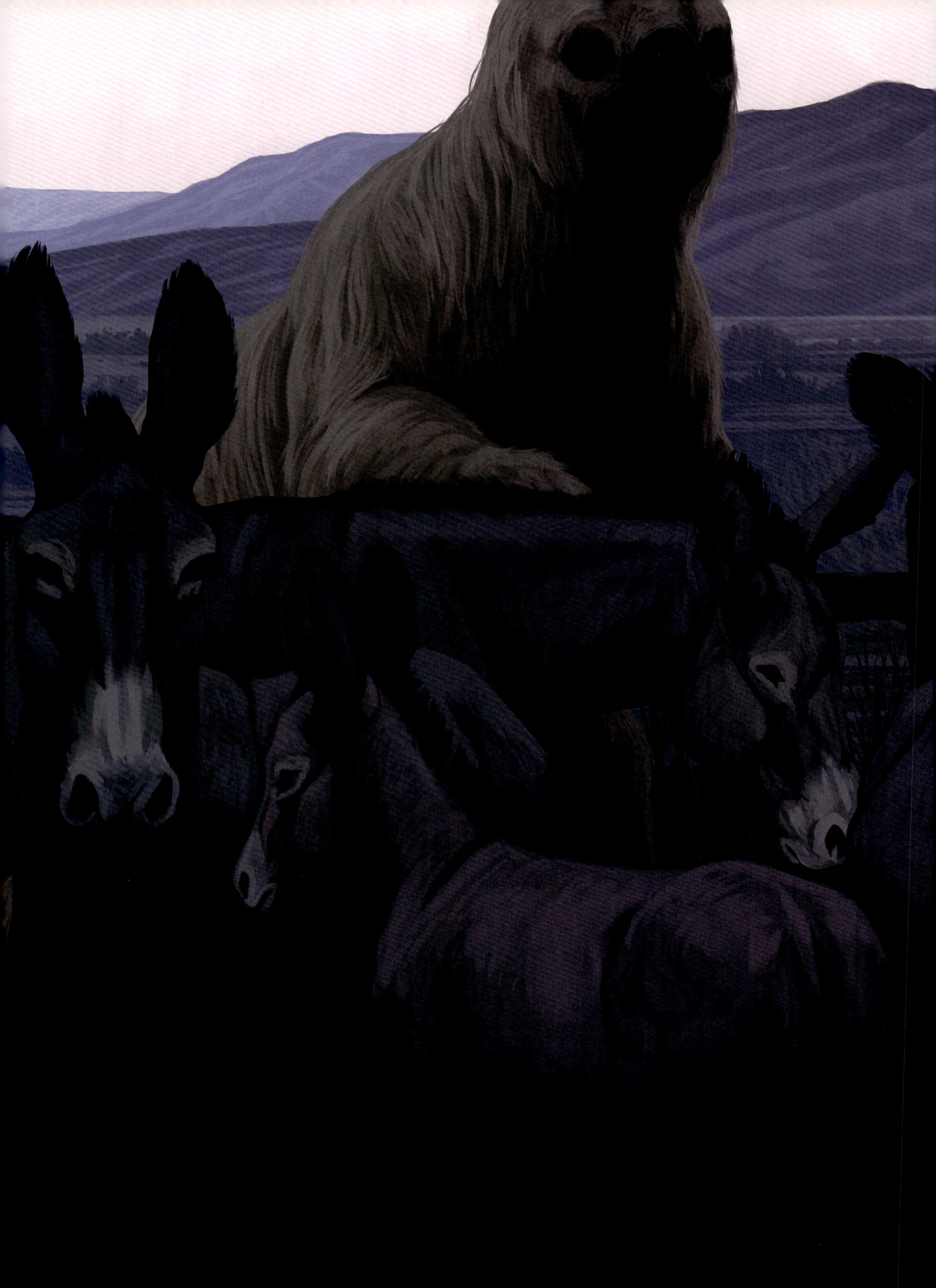

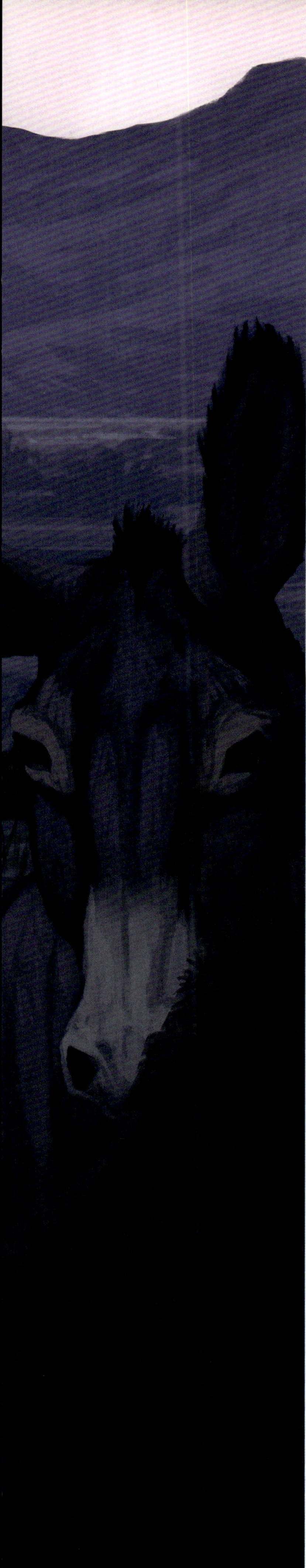

"No idea what it is," Amos said after a few minutes.

We both looked out over the front yard, our vision ending with the floodlight.

"You think we should check? Go out there and see what's making that noise?" I replied.

Amos handed me a beer.

"Hell no. It's not fire, and I don't smell anything like smoke either. It's probably some kind of animal or maybe someone's car. I dunno. I think it's nothing."

"Yeah, true," I said. "I suppose we could just keep watch. Maybe see if anything comes up the road. What an eerie sound that is." I adjusted my back in the deck chair, popped my beer open and took a sip.

Amos nodded and looked down at his hands, holding his beer. He looked upset. "Yep," he said. "I'm not really interested in going out there. Based on what I saw last month, it could be anything."

"What did you see?" I asked.

Amos looked up and out over the front yard. Just a few small cacti and two large Joshua trees were illuminated by the floodlights above. You couldn't see beyond them. Just darkness until the low horizon and the ever-present faint glow of Vegas in the distance.

"This is going to sound weird," he said, then paused and took a sip of his beer as the sizzling sound in the distance grew a little fainter. "But I saw a large creature covered in long red fur. It stood almost upright and had an odd face. It was breathing heavily."

Amos swirled his beer can around and tapped the side with his wedding band.

"There was this stench," he added. "Made me think of that Yucca Man story your dad would talk about. Anyway, this was in the morning. Early. It was standing over behind the donkey pen."

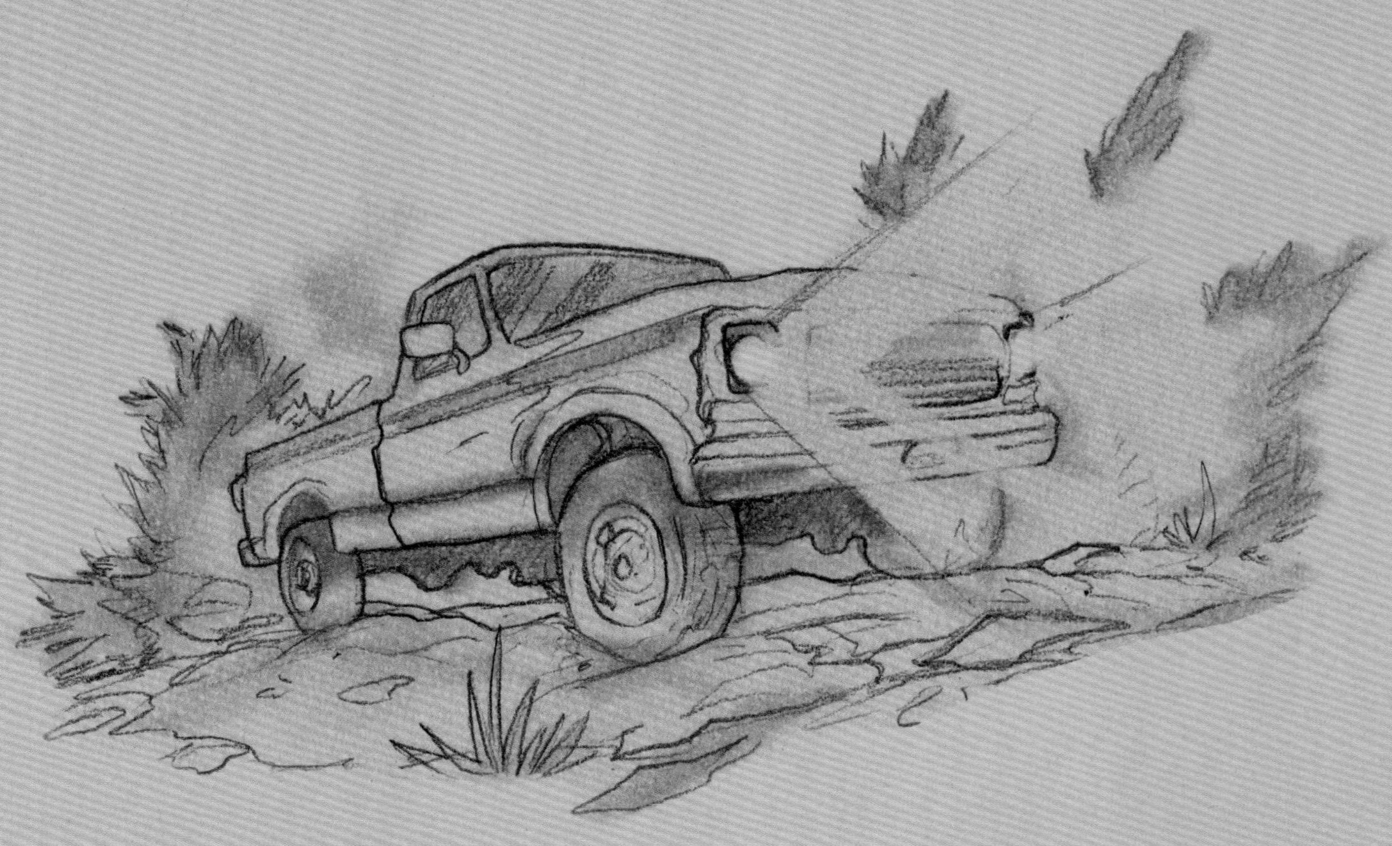

He paused for a moment. "You were still inside with the dogs."

His voice cracked. "Speaking of the dogs, where are they?"

There was a second of silence as we stared at each other. We were both scanning our thoughts for the last time we'd seen Roman and Gayle. We'd just fed everyone, and they were following us, as they always did, as we cleaned and finished up.

Suddenly a cry in the darkness. A dog.

Amos and I didn't say a word. I grabbed the keys from the hook inside. Amos started down the front steps. We bolted to the truck and took off down the fire road. Headed to the south, the sizzling grew louder.

"You see anything?" I shouted at Amos. He leaned over the passenger-side window, scanning the dark desert for any sign of Roman or Gayle.

"Goddamnit! No!" he replied. "Maybe they're back at the house? What if it was just a coyote or something?"

The road down here gets rough, and the truck bounced as I swerved to avoid a large rabbit hole.

The sizzling sound was all-consuming. I could barely hear anything outside of it. "That wasn't a coyote!" I shouted back. "They don't cry out like that! It sounded like Gayle to me!"

Amos looked ahead. "Whoa whoa whoa!" he shouted, unable to form words.

I hit the brakes, bringing the truck to a sudden stop. Everything moved forward, then back. Dust filled the air, and the sizzling grew louder.

The juniper trees were lit up by the truck headlights. Just behind them we could make out what looked like moving stones: A mass of small red bodies hopped and writhed over each other in a tangle of meaty legs and chubby stomachs. Eyes reflected like shimmers over a creek. The sizzling was all we could hear.

"Toads!" I shouted. "Red toads!"

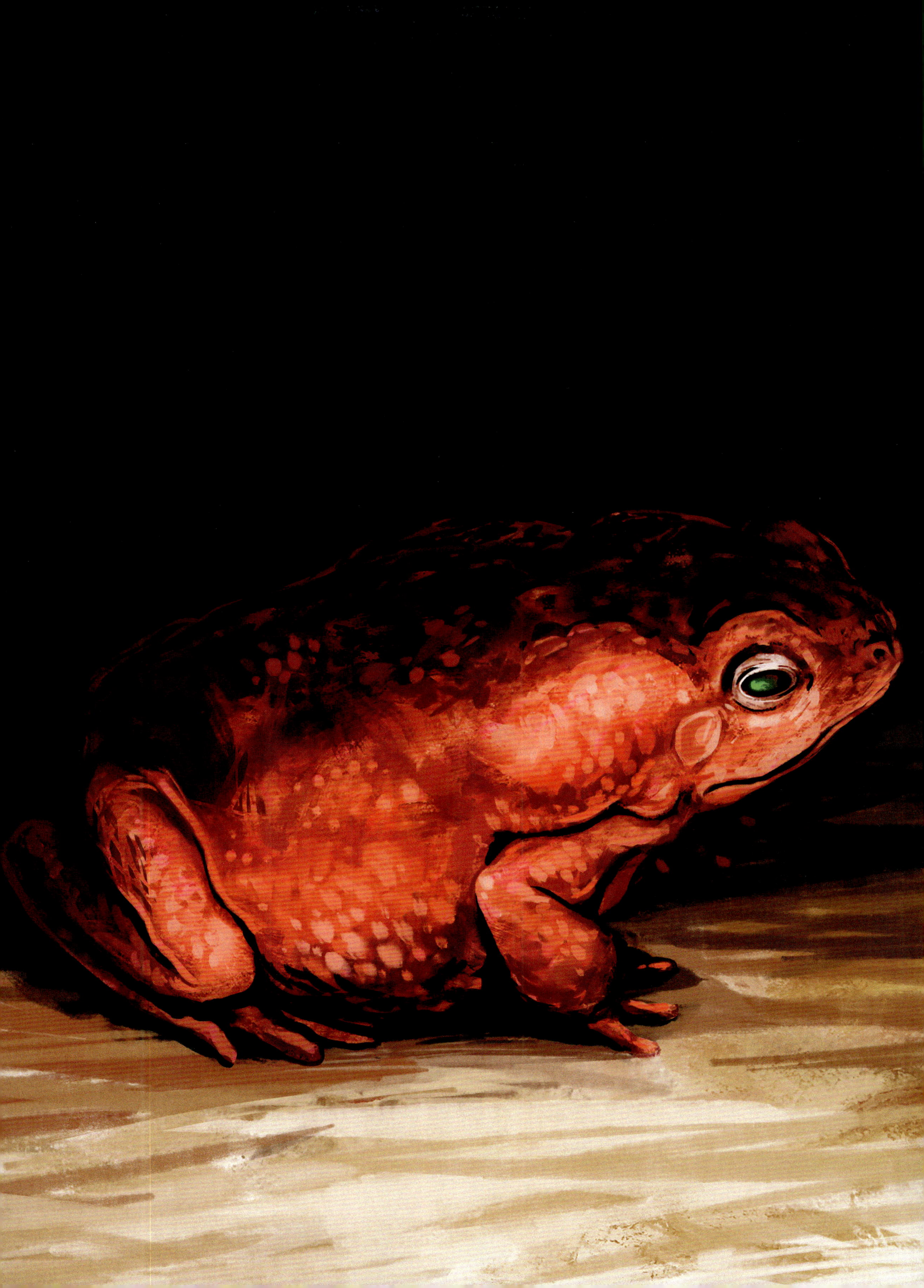

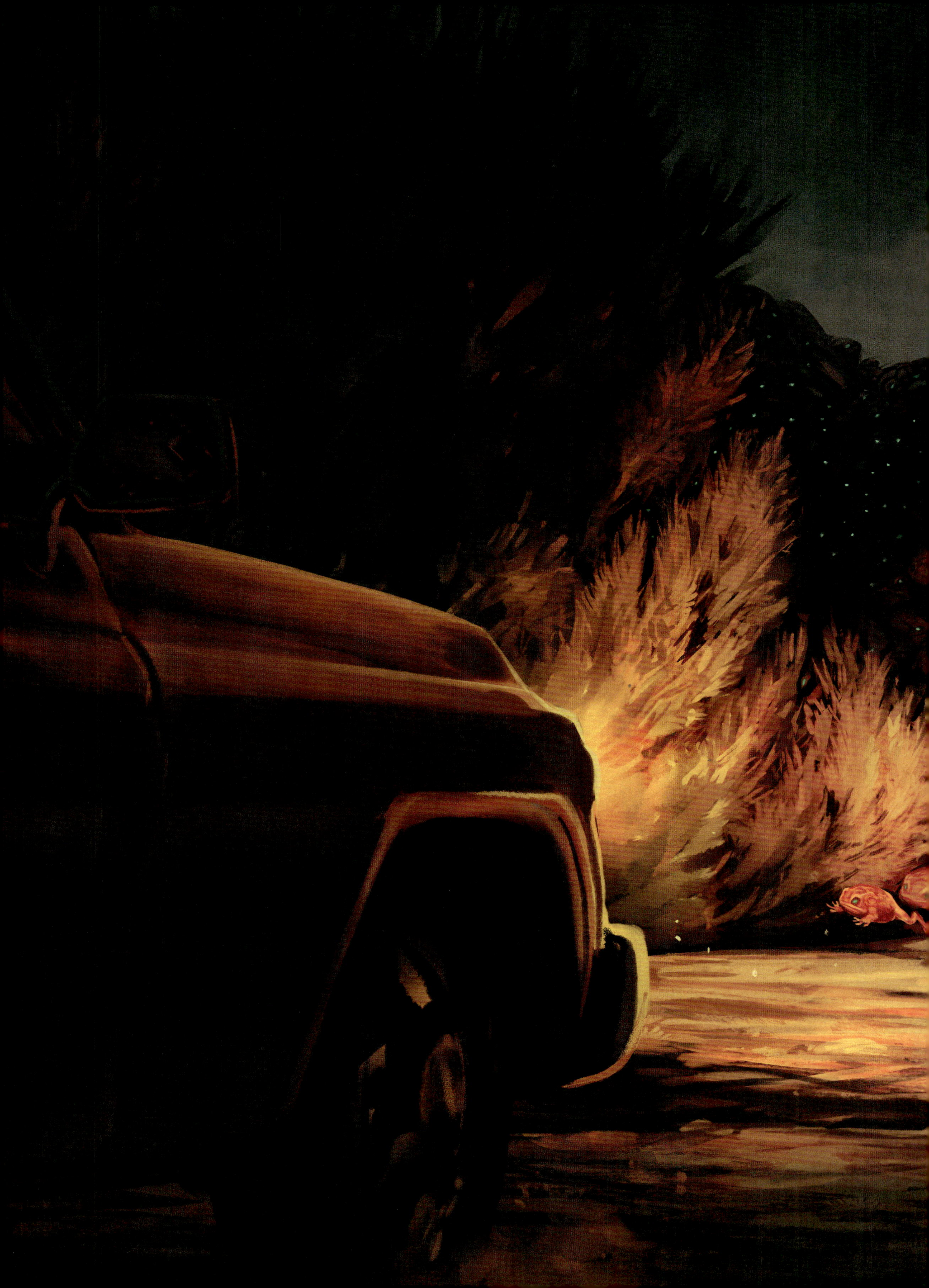

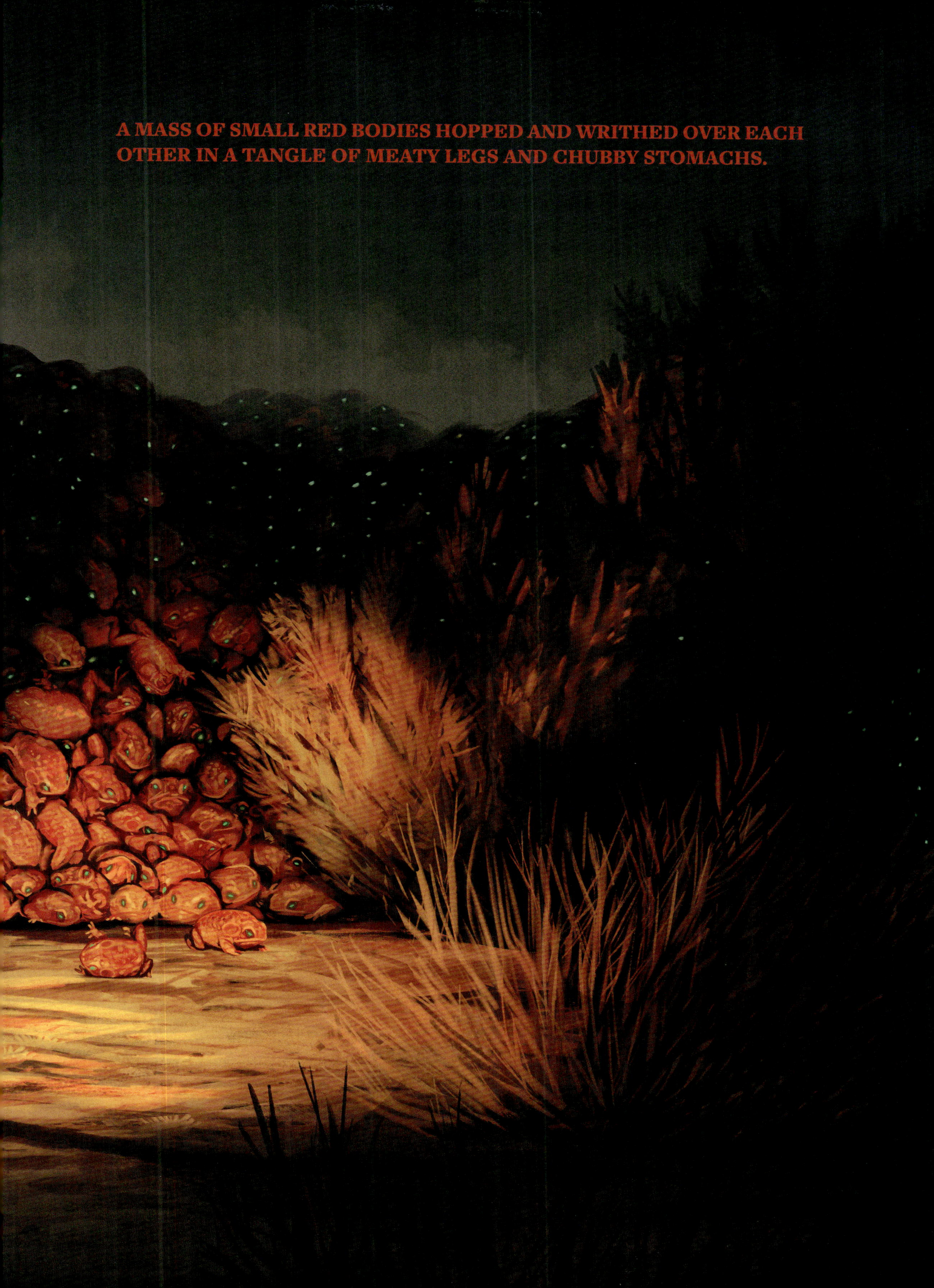
A MASS OF SMALL RED BODIES HOPPED AND WRITHED OVER EACH OTHER IN A TANGLE OF MEATY LEGS AND CHUBBY STOMACHS.

The dust around the truck started to settle, and the toads moved as an entangled mass all around the bushes and rocks—slapping each other with their moist bodies, pulling themselves over and under one another. Suddenly the color of everything became warmer, red and orange lining the edges of things. The junipers were glowing.

We got out of the truck and stood next to the open doors, engine running and lights still ablaze.

"Gayle! Roman!" I could barely hear Amos' voice shouting into the swarm.

The toads clumsily puddled over one another. I would focus on one only to have it covered up by ten more. Some were dead along the sides of the swarm. Eyes open and lifeless, their skin a deep burnt red. They were like clay monsters brought to life. Their sound was deafening.

I looked at Amos. He hadn't stopped shouting for the dogs. We were both in tears. Dust had collected on our cheeks as mud.

He looked at me.

We got back into the truck. I put it in reverse and backed away down the road. The toads receded into the darkness, their noise still filling the air.

I backed up into a small turnout, stopped and looked over at Amos. He was crying and angry.

"What the hell is wrong with this place?" he said.

He stared into the darkness, the headlights illuminating the dust and desert brush near the road. The truck engine's sound barely met the level of the toads.

"I'm sorry, dear," I said. "I'm so sorry. We'd better head back. Maybe the dogs made it home." I felt

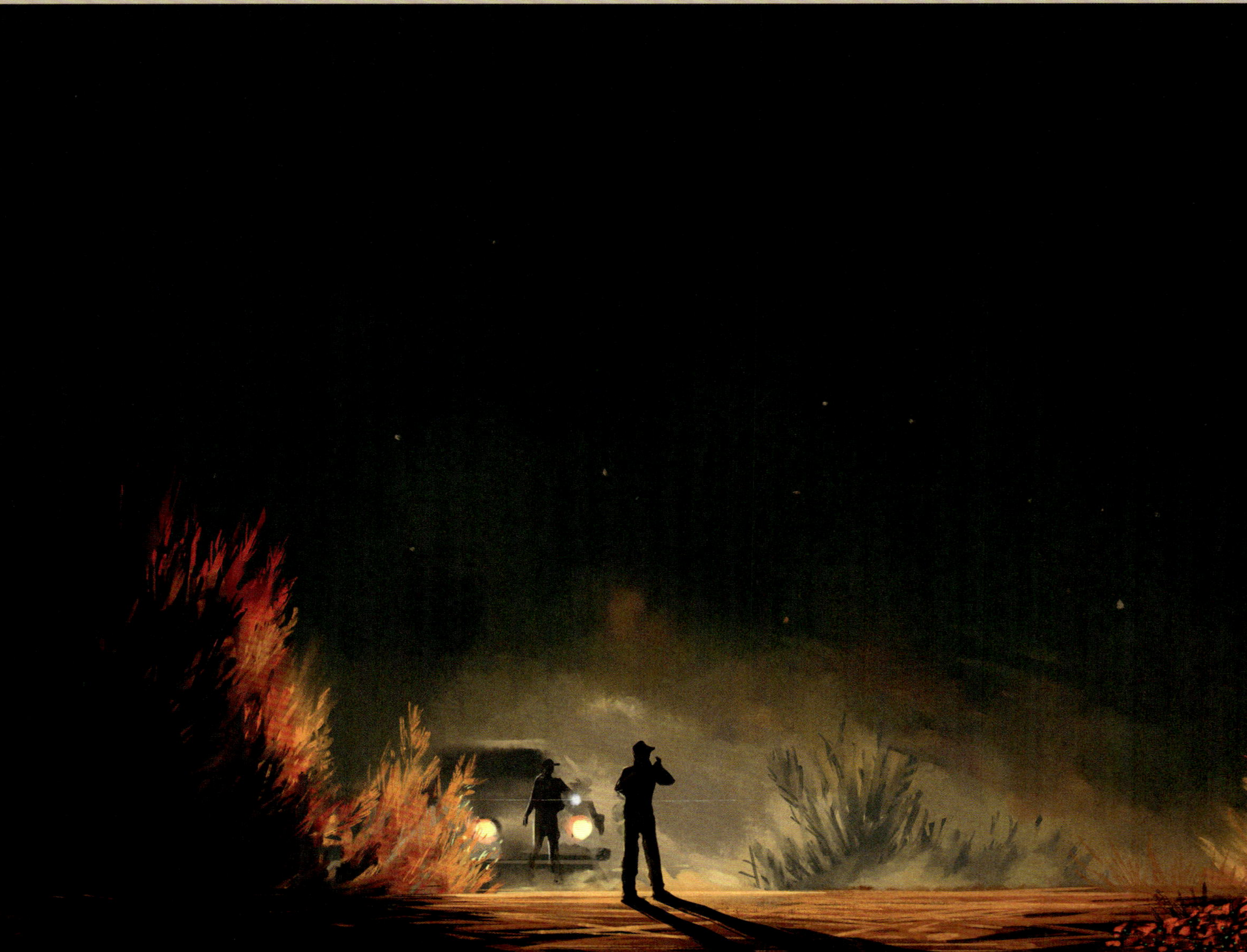

hopeless, but I wanted to tell him something.

Amos sat in silence. He closed his eyes and took a deep breath.

I turned the wheel of the truck and slowly pushed down on the gas.

We drove back in silence, both of us trying not to look too hard into the darkness, both of us trying to keep it together. I kept picturing Gayle with her long legs and her white tail tip, running through the brush.

We made it back. As we got out of the truck, the sizzling sound had subsided a bit, but it was still there—hundreds of those toads out there in the night. Tangled.

We walked up the porch steps and picked up the spilled cans of beer. We walked through the house, calling for the dogs. Nothing.

Amos circled the house a few times. I stood and looked down the fire road we had come from, thinking I'd spot our pair of dogs trotting up playfully as if nothing had happened—like kids who knew that they had been out after curfew.

It was pitch-black now. Had to have been midnight or so. A new moon was out. No light except the floodlights out front. We glanced at each other and made our way to the porch, the sizzling sound still pulsing out in the dark. Amos walked over to the cooler and grabbed two beers. We sat there listening for the dogs, concerned but numb. Drinking in the dark.

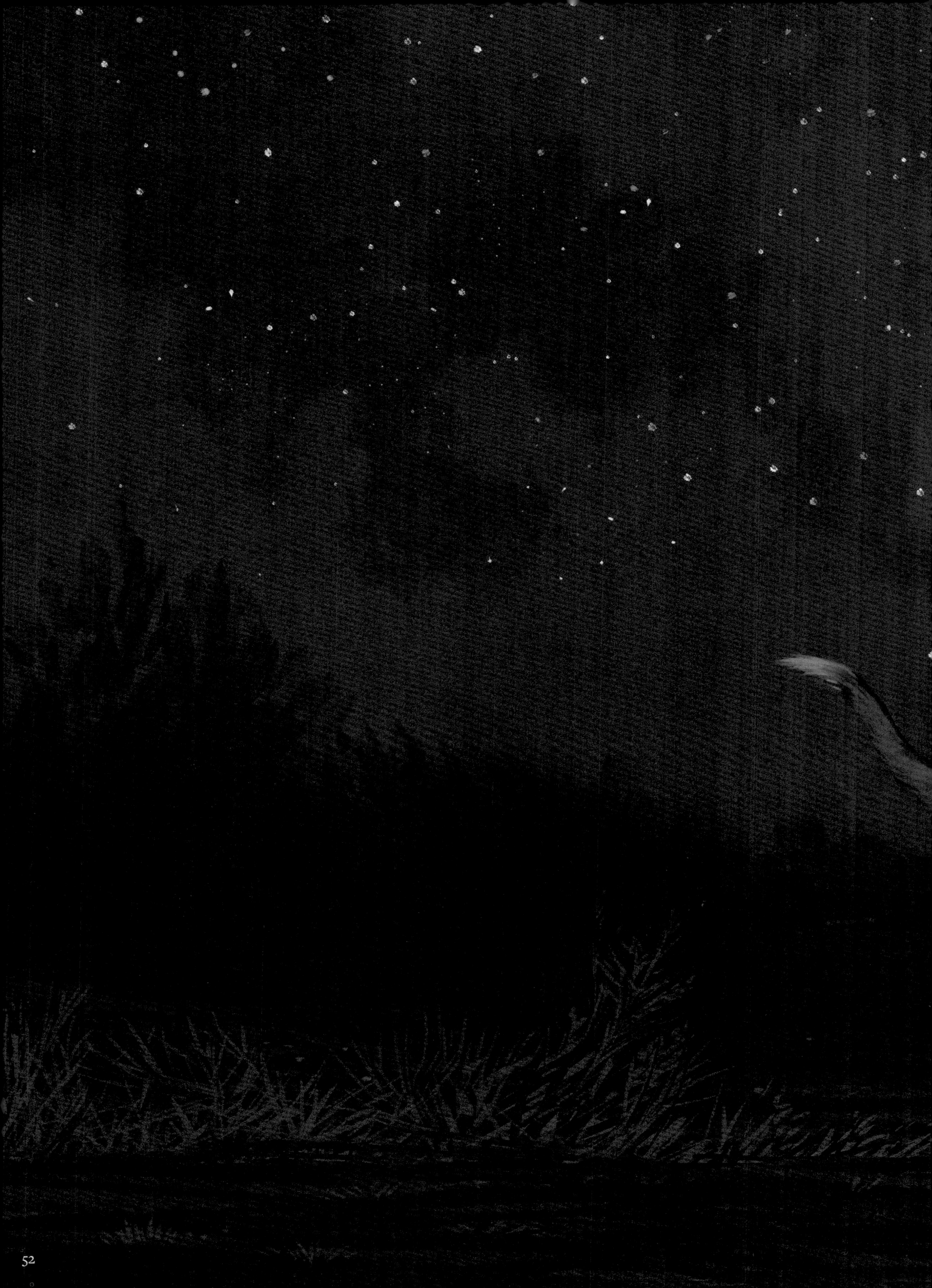

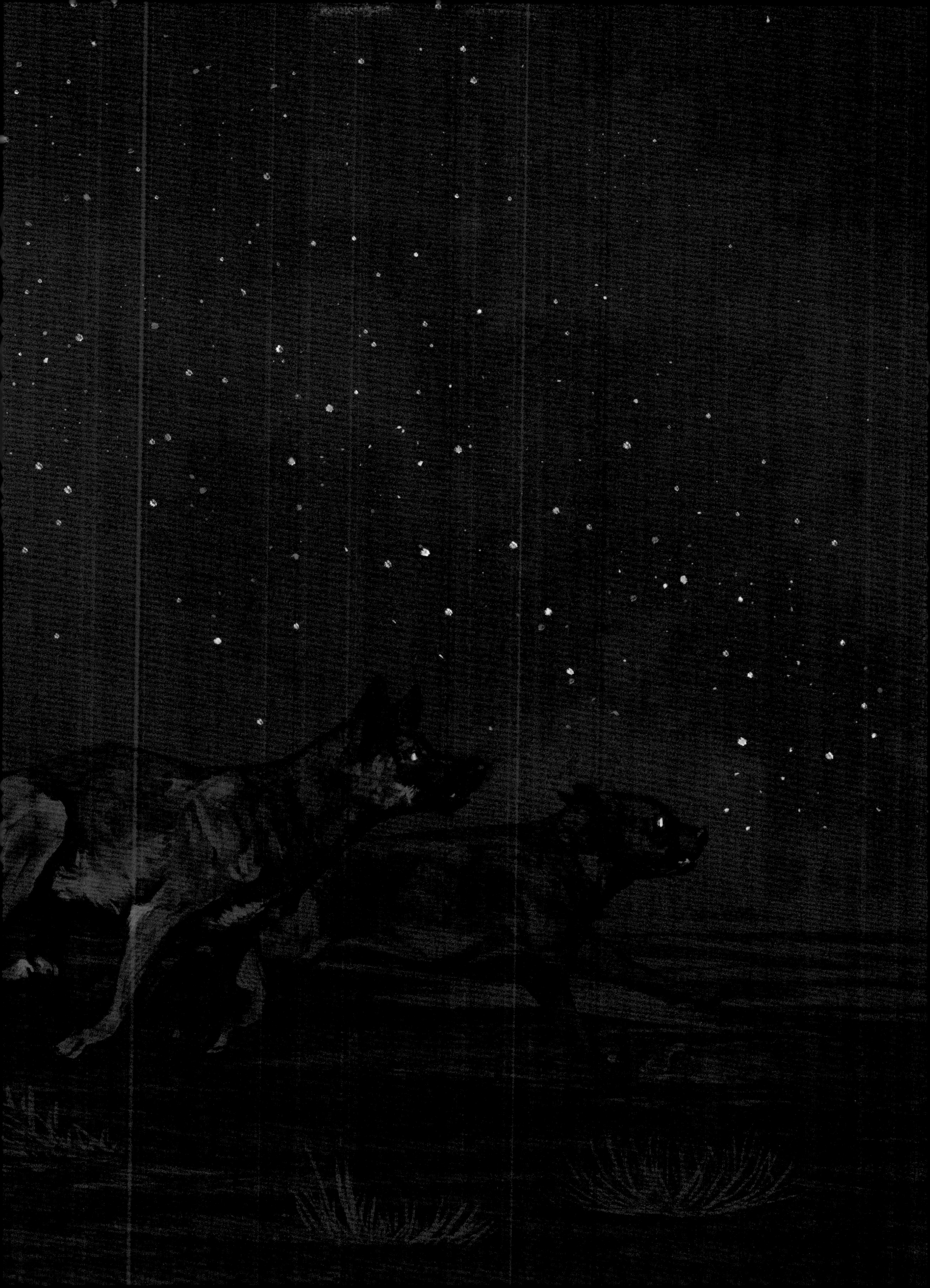

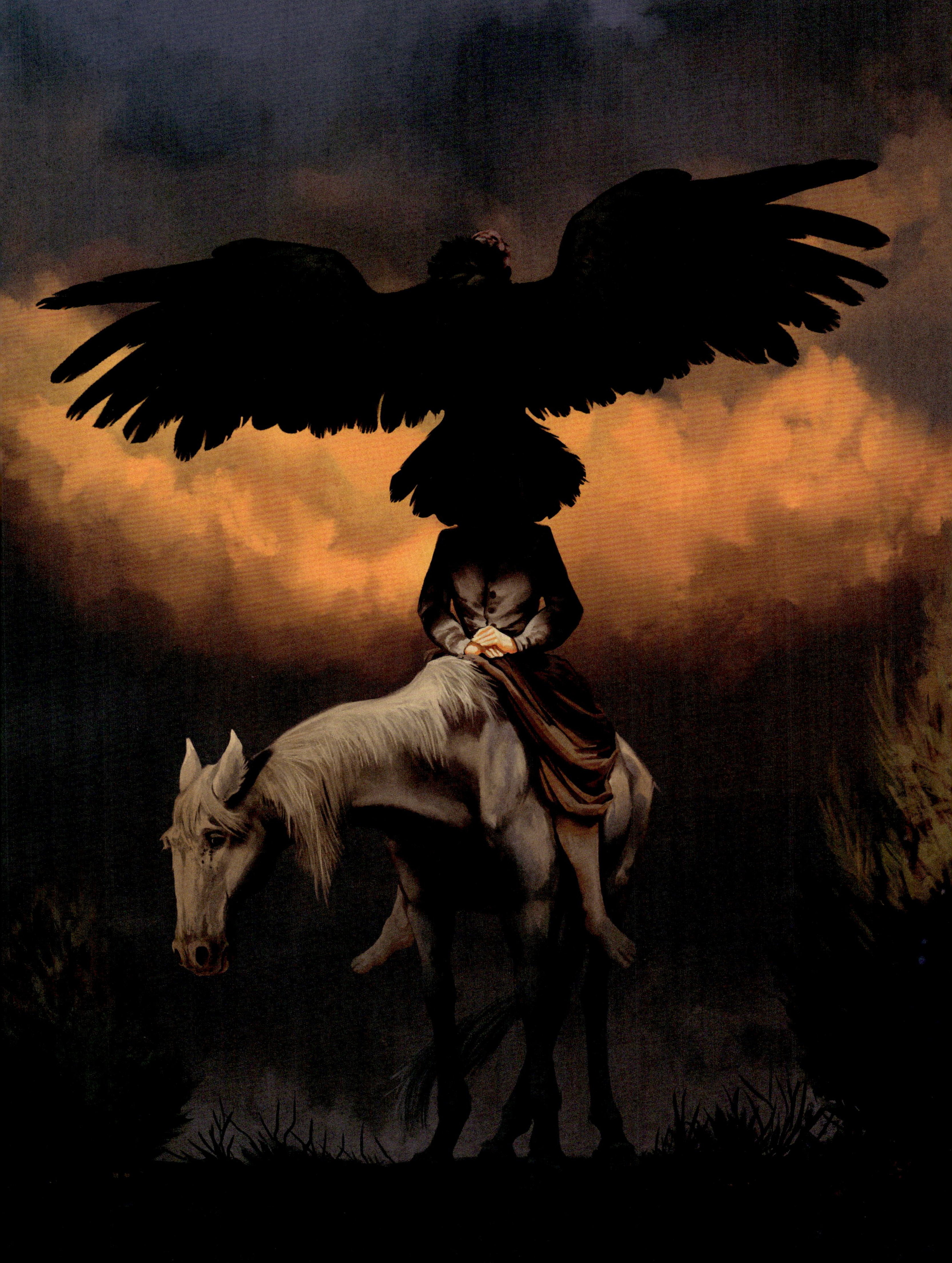

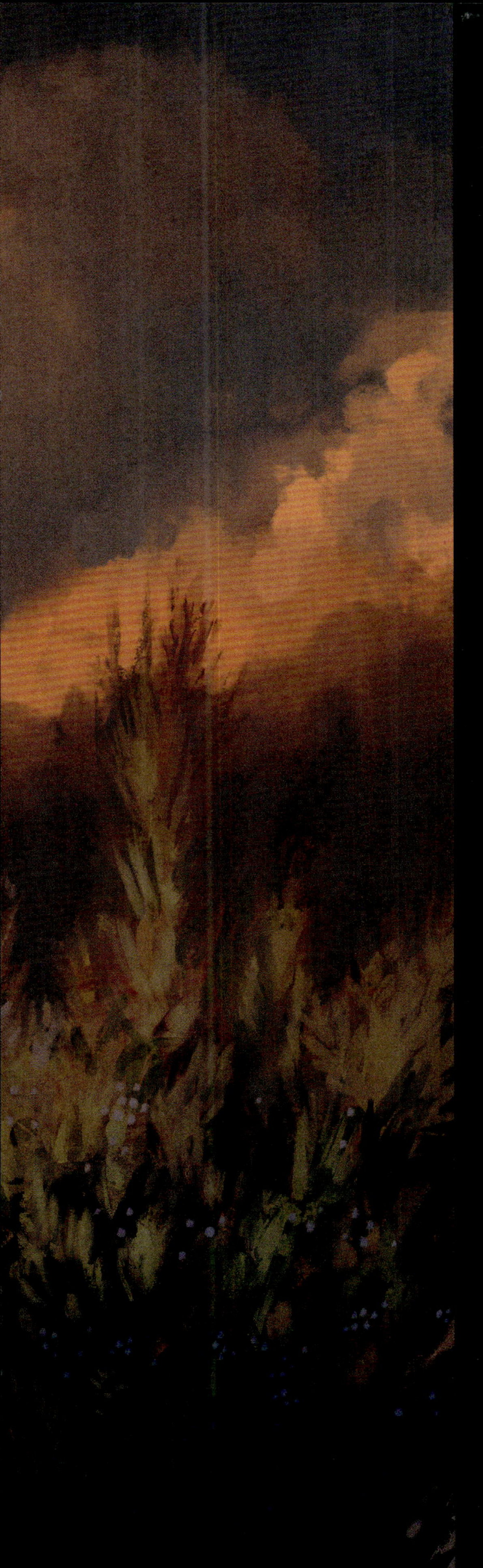

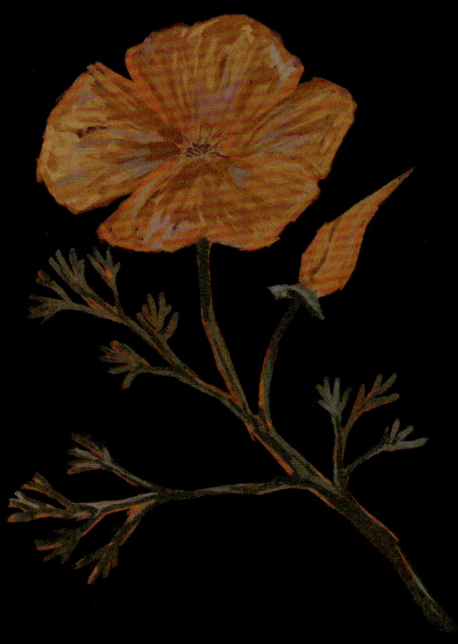

THE PIONEER WOMAN

Location: Omitted.

Considerations: This is an old ghost.

She sat silent under an angry sky.
    Among the junipers and early wildflowers of spring.
    A harbinger of destruction.

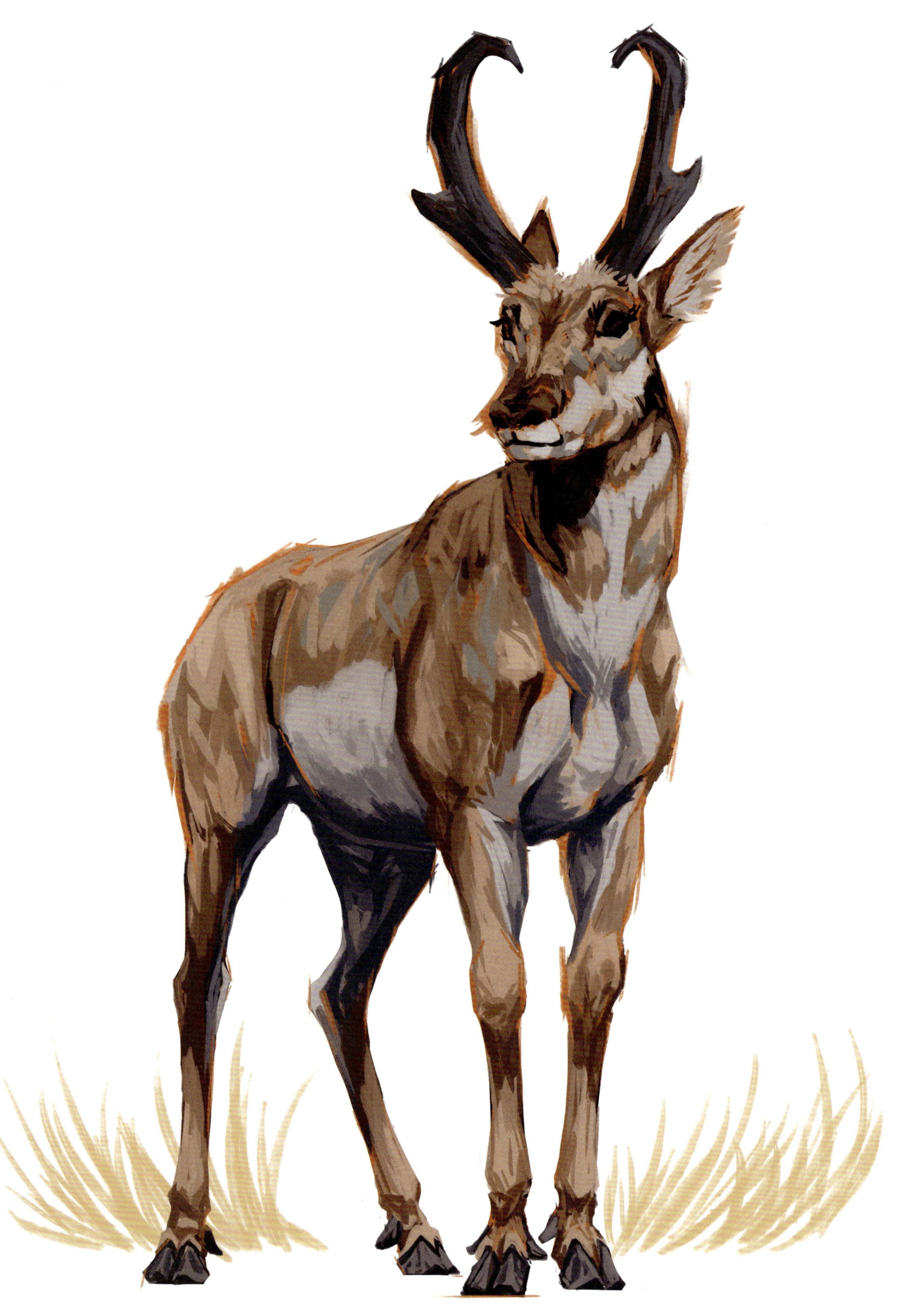

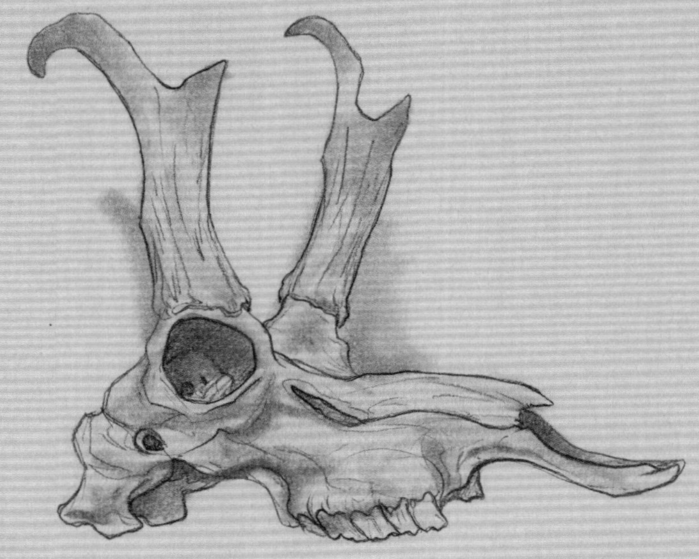

PRONGHORN ANTELOPE

Location: The pronghorn used to range all the way from Canada to Mexico. They still reside in broken populations today.

Considerations: These animals are not related to antelope but share a common ancestor with the giraffe.

The fastest mammal in the American Southwest, the pronghorn antelope evolved to outrun the now-extinct North American cheetah. Today, it barely keeps up with the sprawl. Such an ancient-looking animal. It hearkens back to an older Earth.

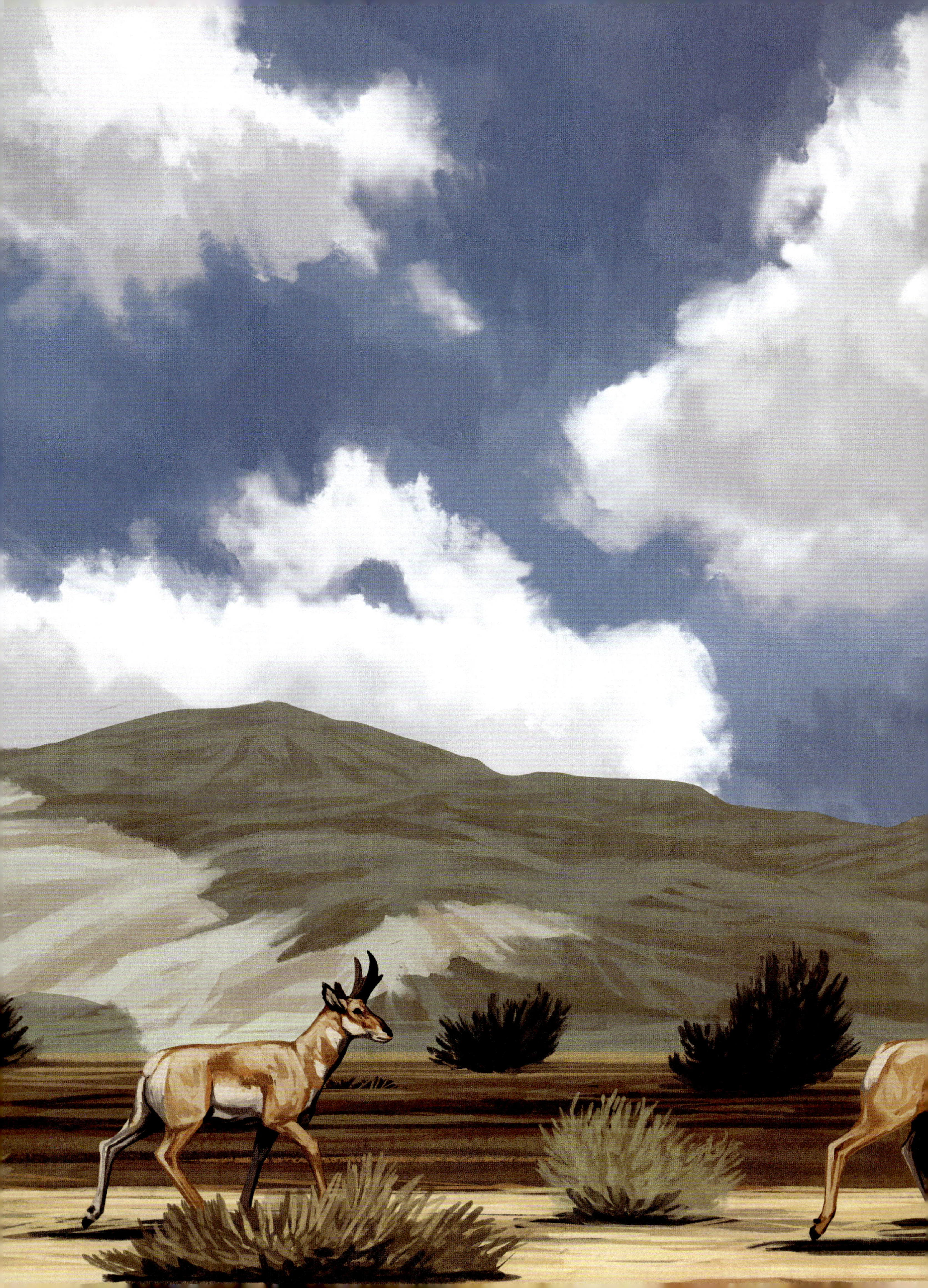

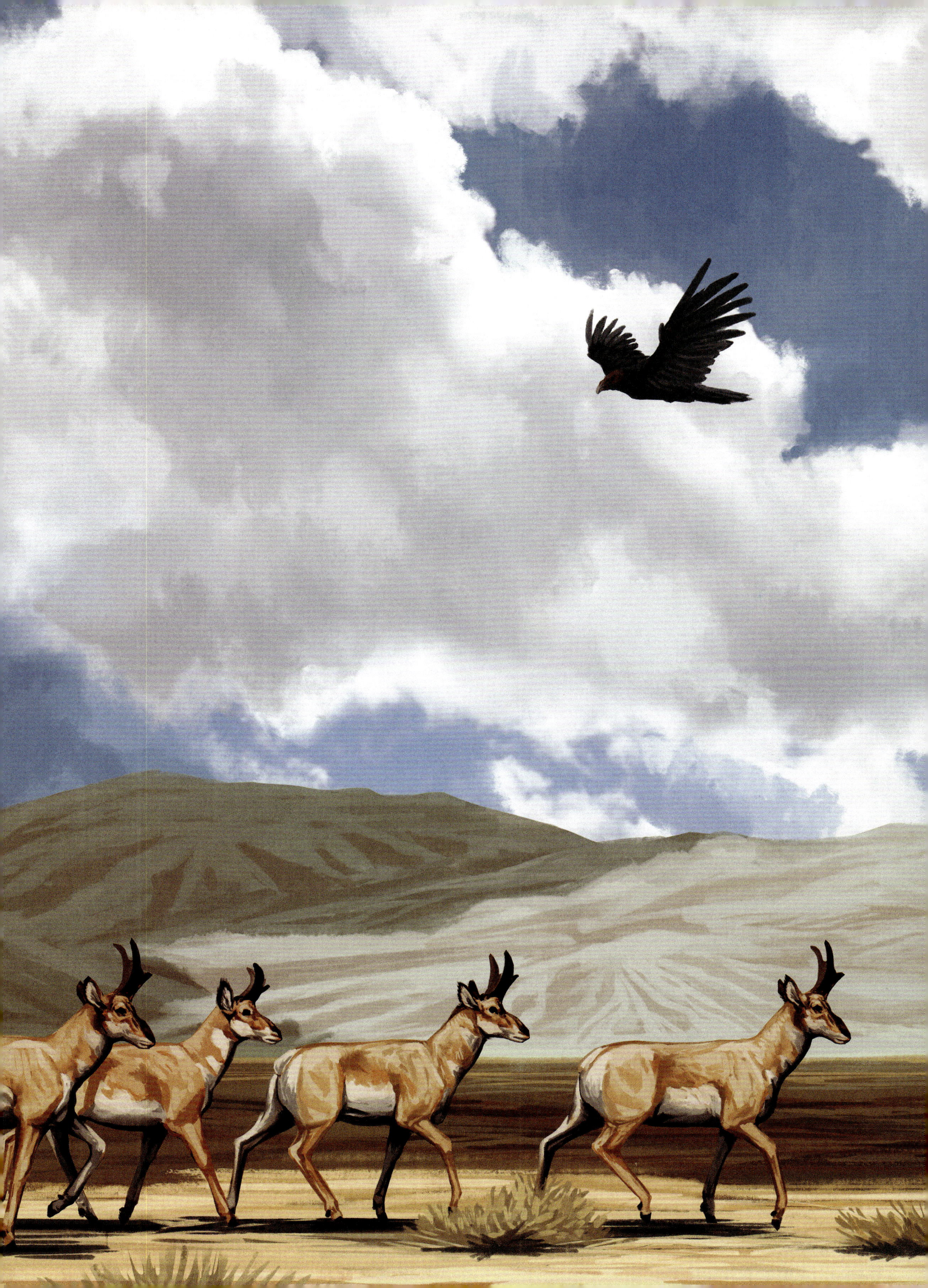

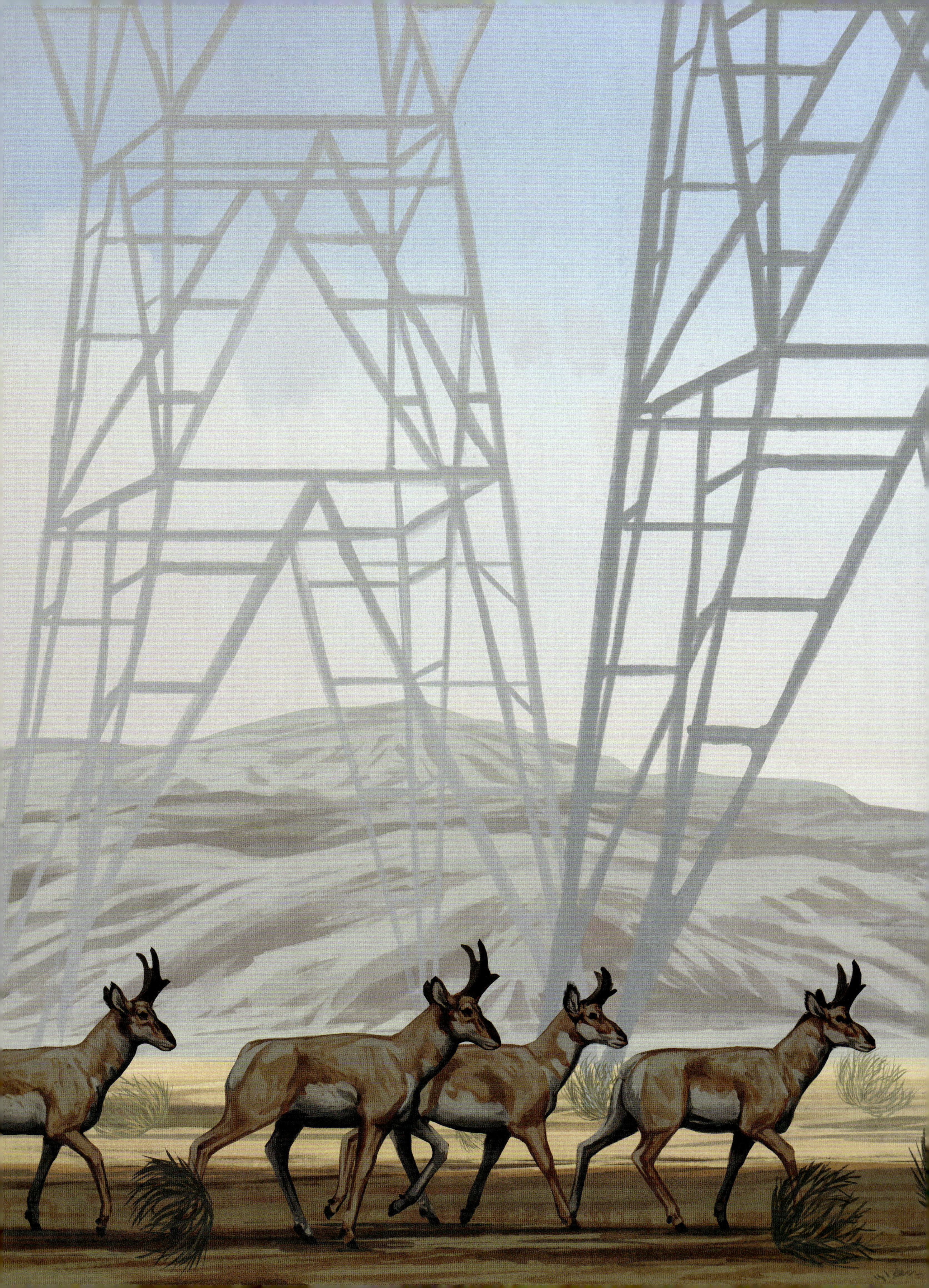

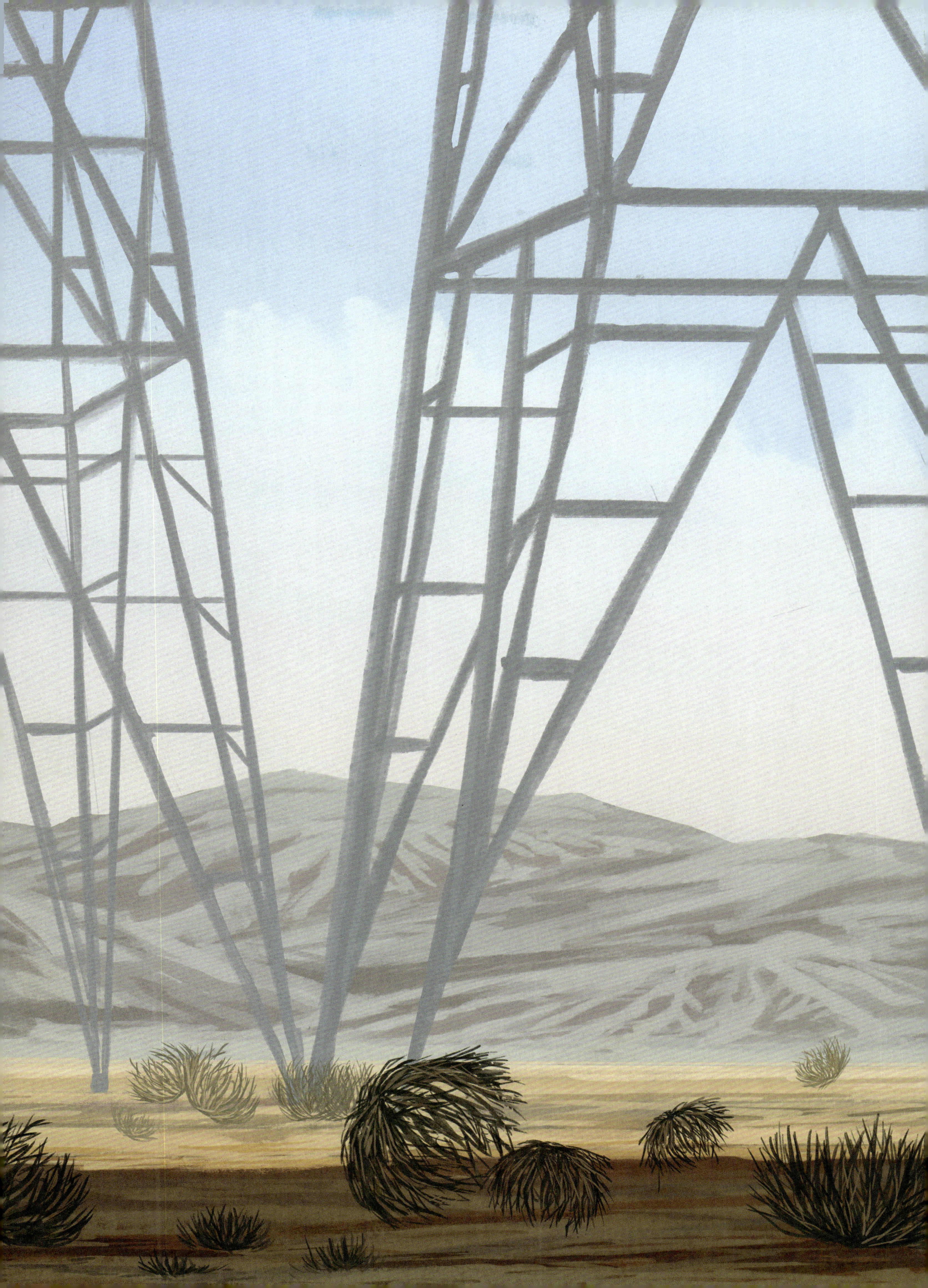

GHOST DOG

Location: East Mojave, out near Saddleback Butte.

Considerations: Don't follow this ghost. She will lead you to trouble.

Most small and forgotten towns of the American Southwest have stray dogs that wander the streets. Some are ghosts.

There's a dog that can be seen walking along the roads in east Mojave. I've encountered her many times, ambling among the garbage. Her coat absorbs the sunlight. Her eyes are hollow. What would be a pink tongue is pale, dead.

She typically roams alone, and she disappears almost as instantly as she comes.

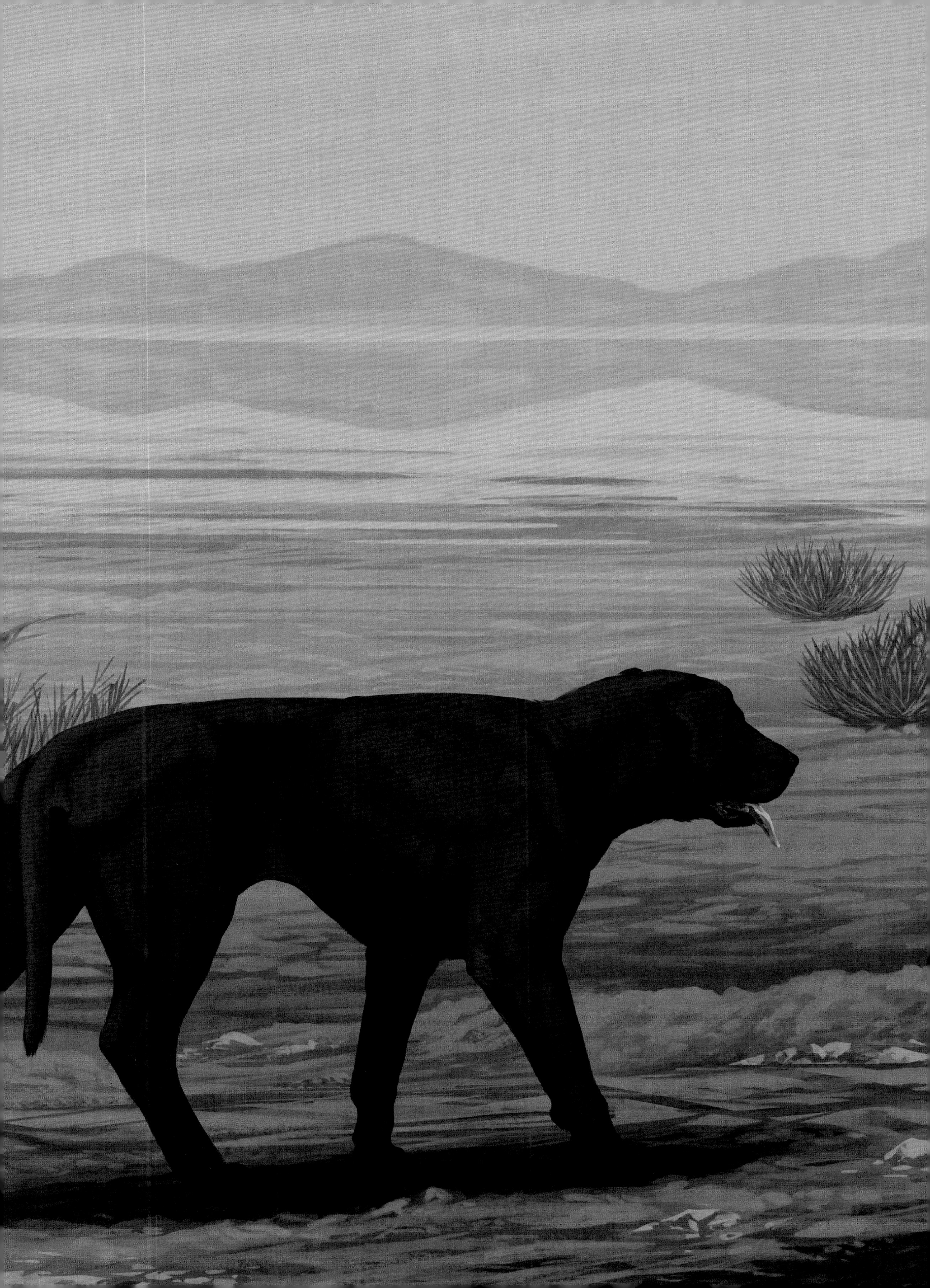

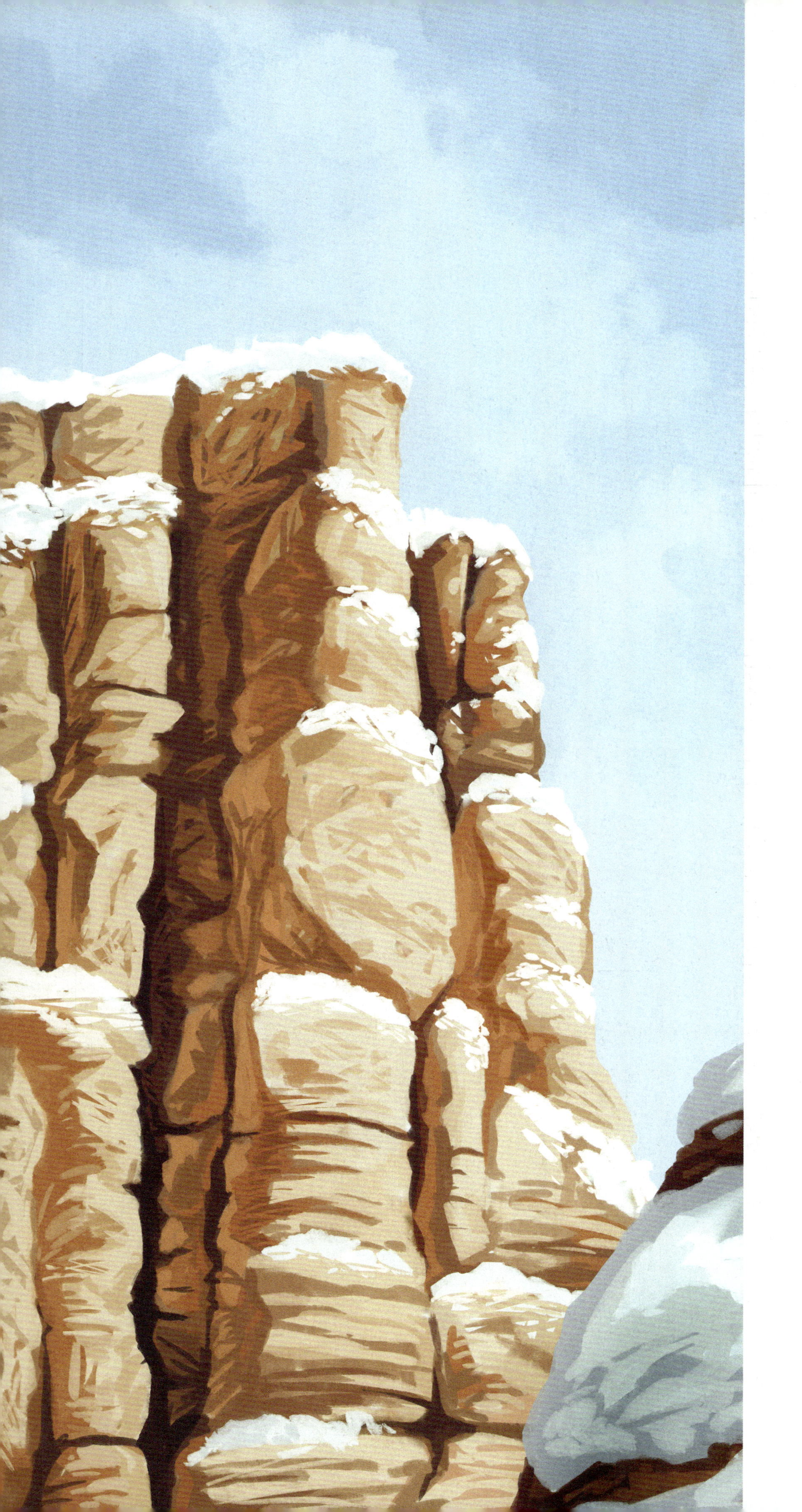

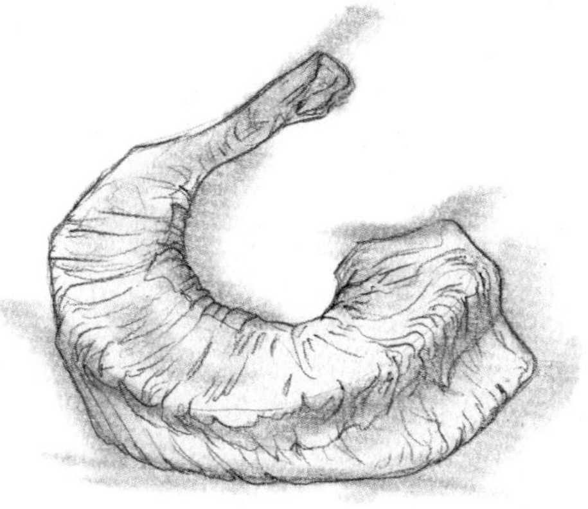

THE TANGLE

Location: Southern Utah.

Considerations: High desert winters don't make you sweat, but you'll be thirsty as hell. Bring water and warm clothing.

In the white winters of Utah, a violent mass of many horns is said to be seen roving on the outskirts of town. Stay away from this one, friends. It's angry.

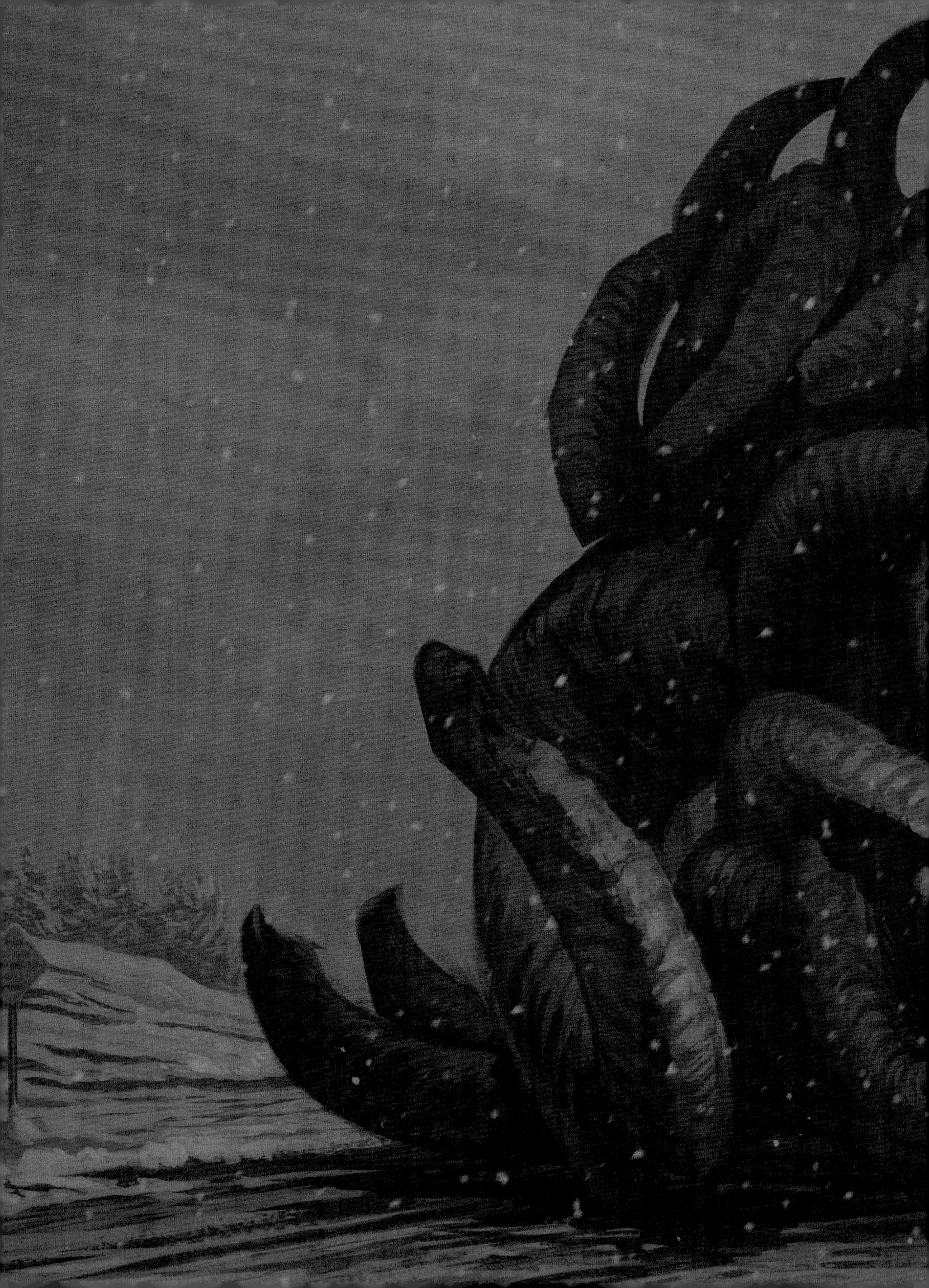

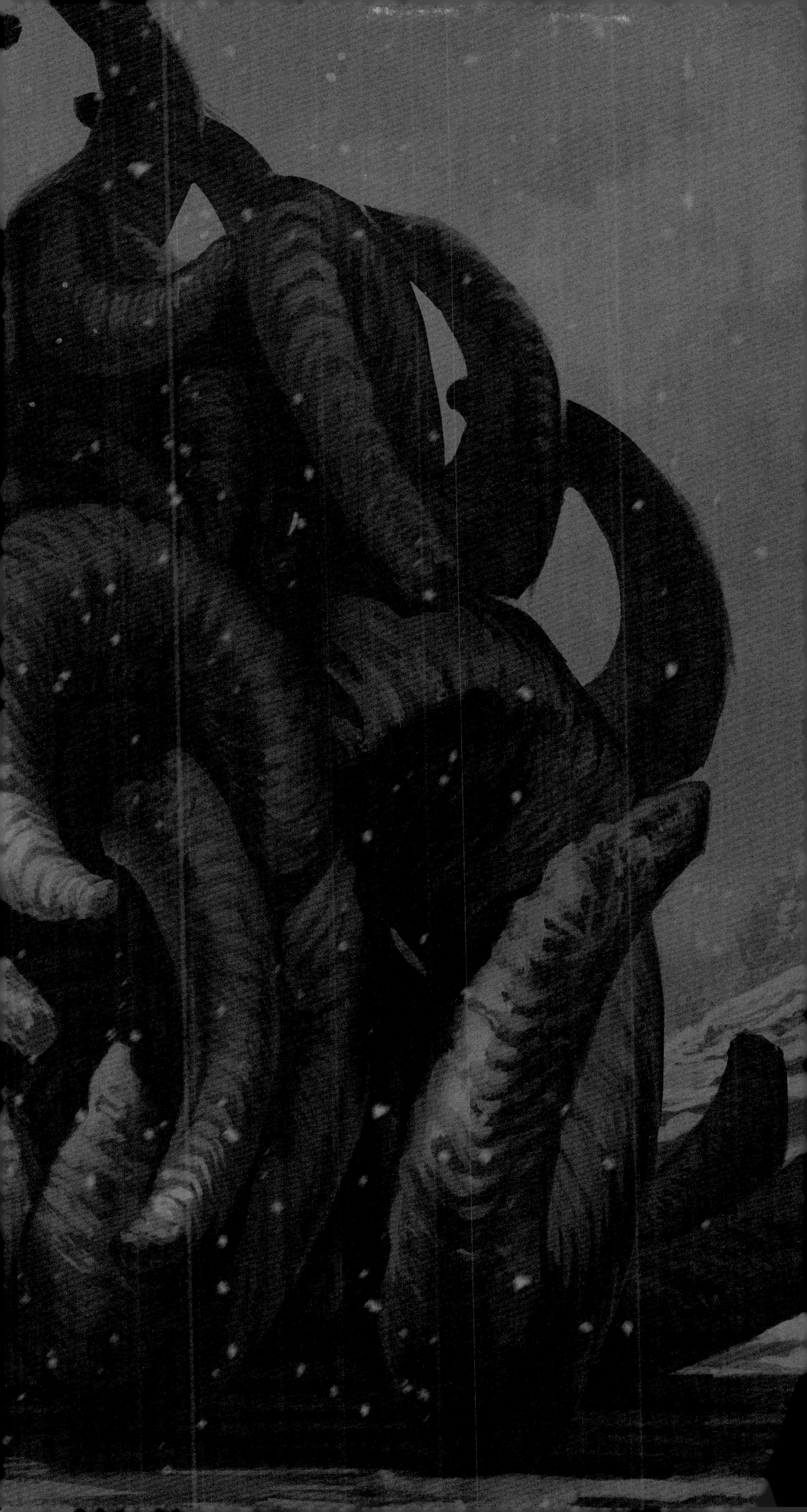

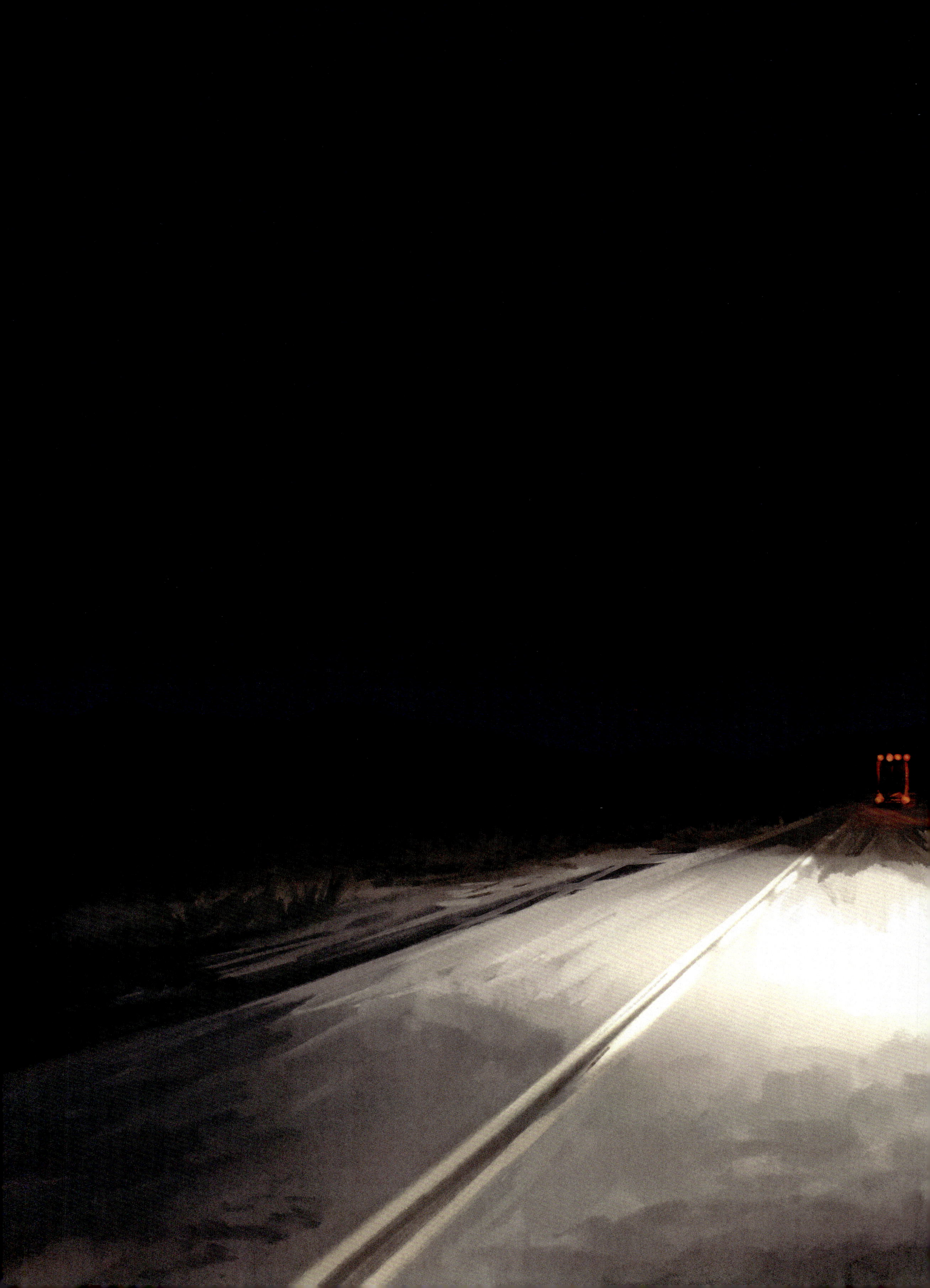

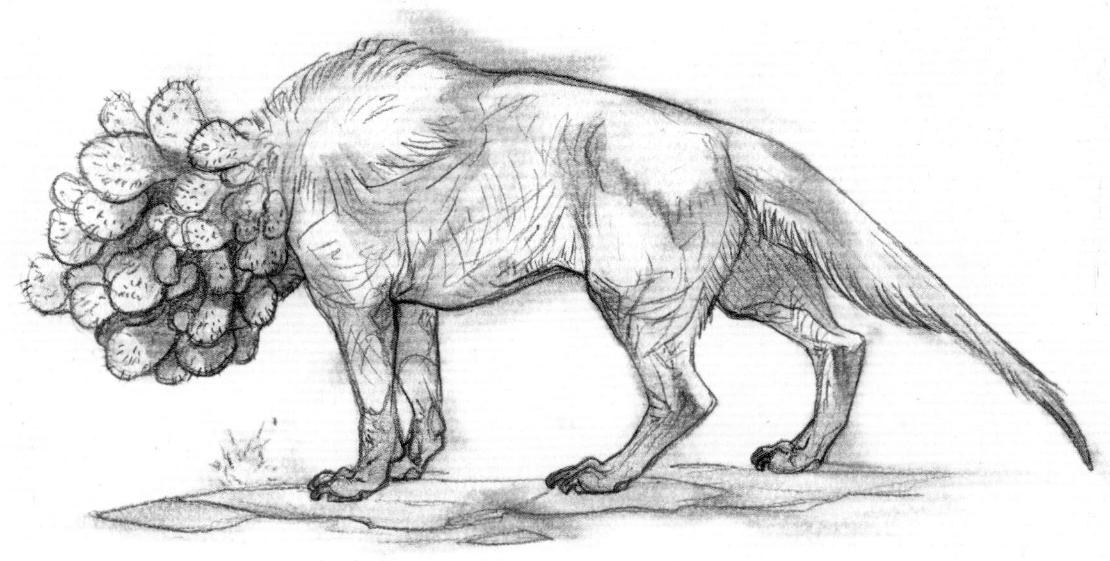

CACTUS HEAD

Location: Mojave Desert.

Considerations: Nobody knows if Cactus Head is dangerous or not. I wouldn't test it.

A lesser-known cryptid of the Mojave Desert, this elusive creature goes by the name of Cactus Head. A few sightings of the thing have been recorded in the Yucca Valley and Joshua Tree area. Cactus Head has been described as a large canid-like animal with a long slender tail and a bushy cactus for a head. It's been seen mostly by semi-truck drivers, trotting across dark highways and alongside their rigs as they drive. One man by the name of Earl Reymund shared his account of the beast, cornering him as he walked from a rest stop to his truck.

"I was walking back to my truck, and the thing came out of nowhere. Stood there right in front of me. I didn't think about it too long. I just yelled at it, 'Hey, bear!' It was all I could think of at the time. Anyway, I think that spooked it."

—Earl Reymund
Southern Nevada, Interstate 15

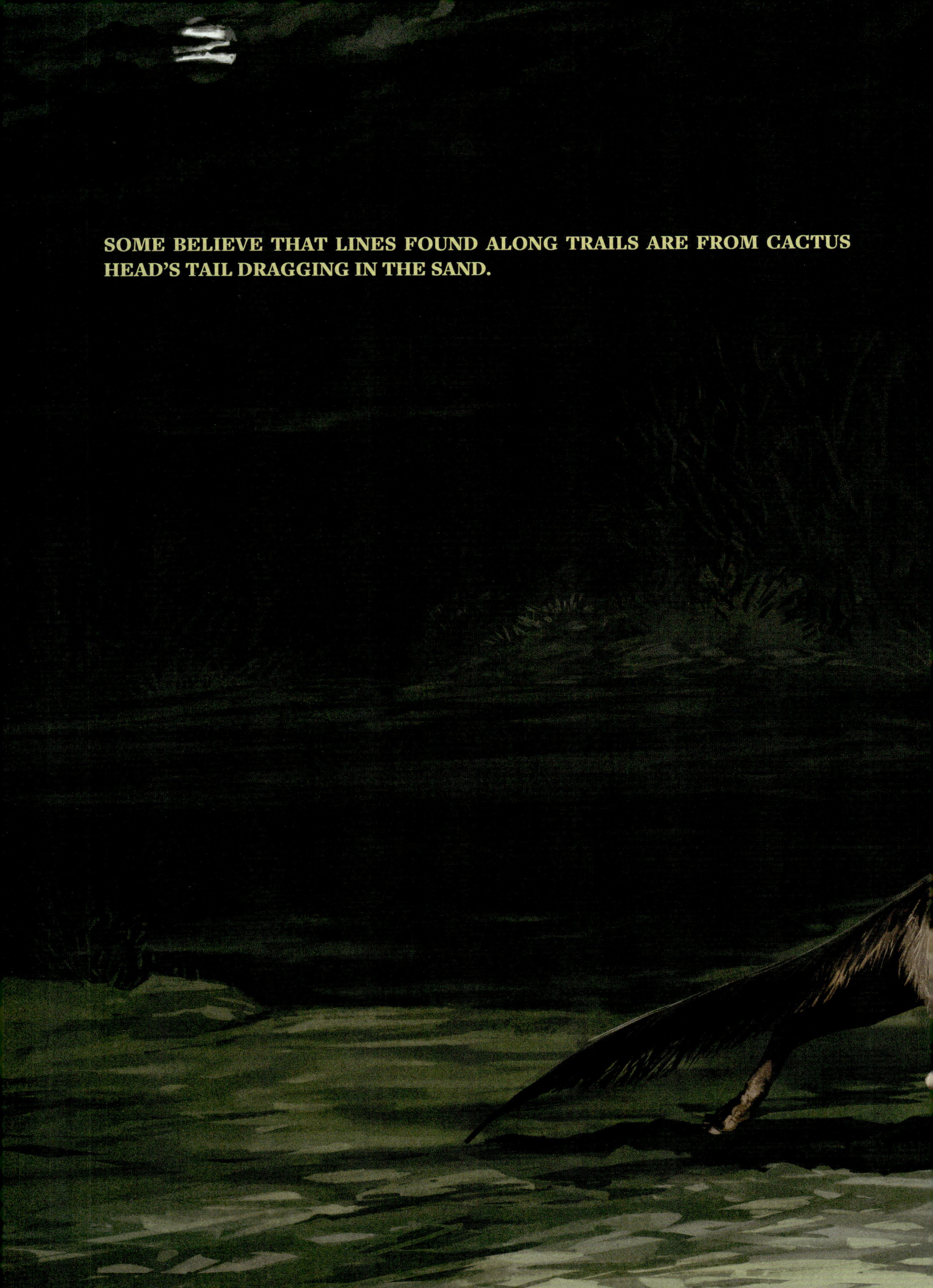
SOME BELIEVE THAT LINES FOUND ALONG TRAILS ARE FROM CACTUS HEAD'S TAIL DRAGGING IN THE SAND.

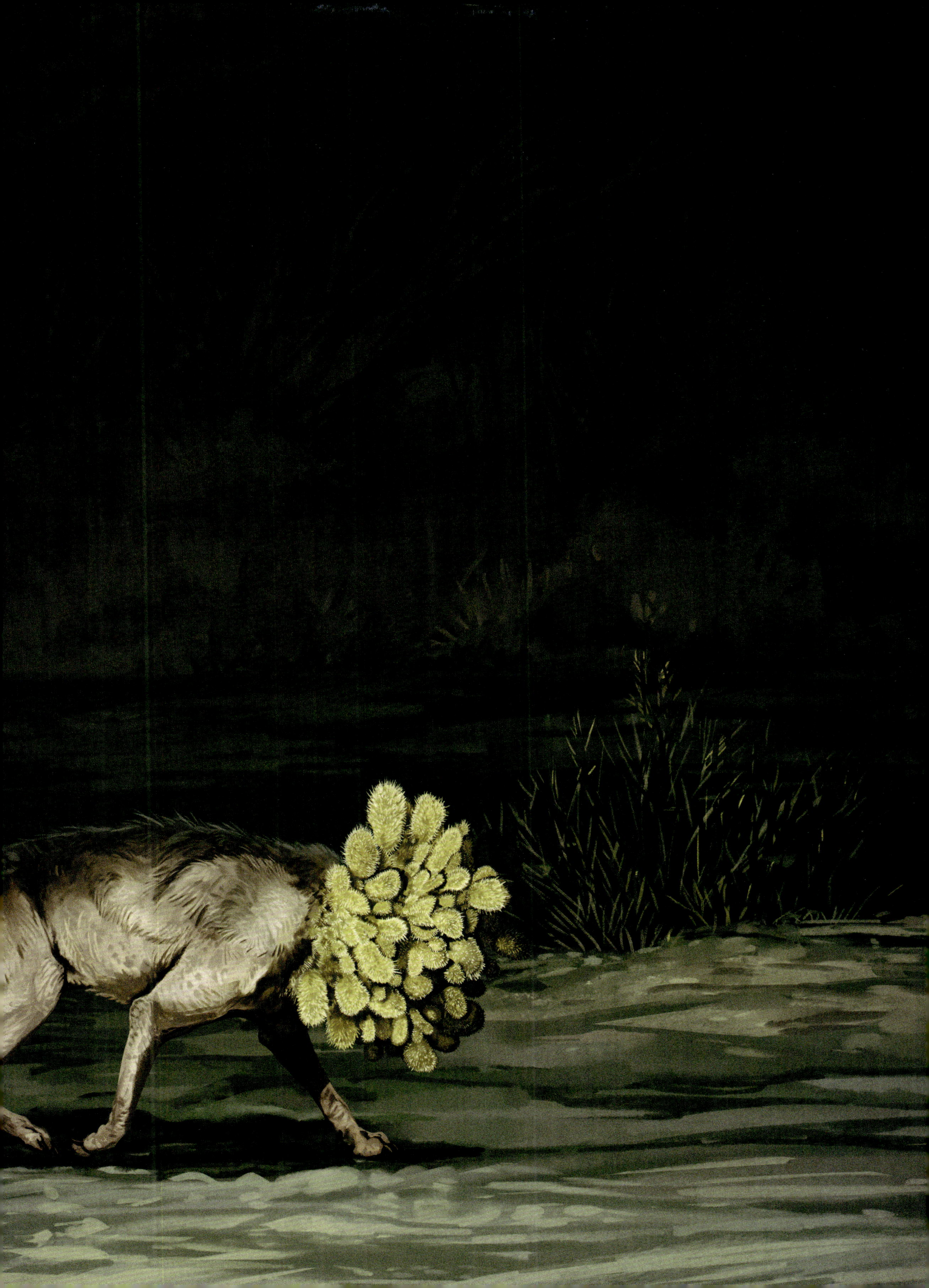

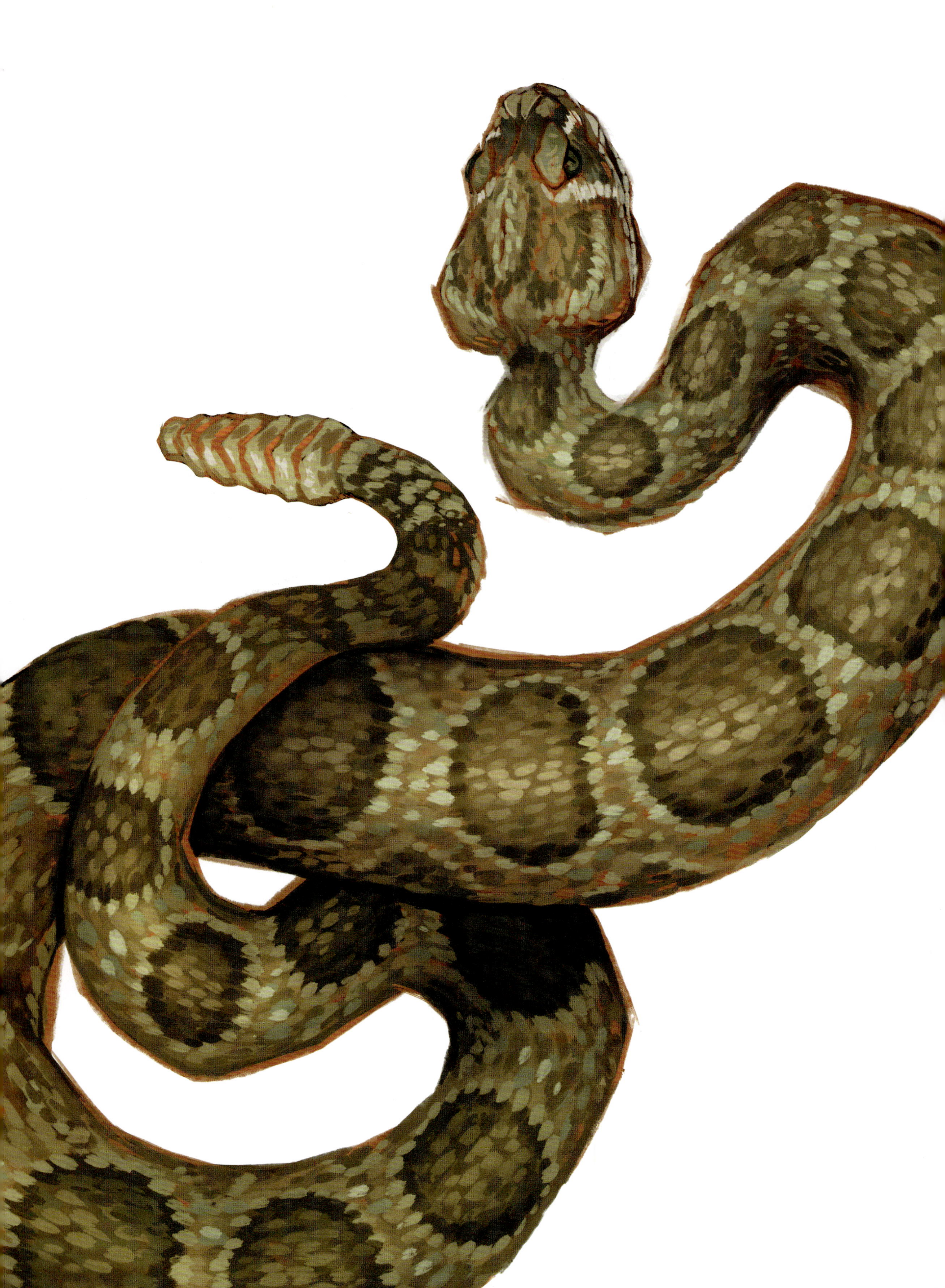

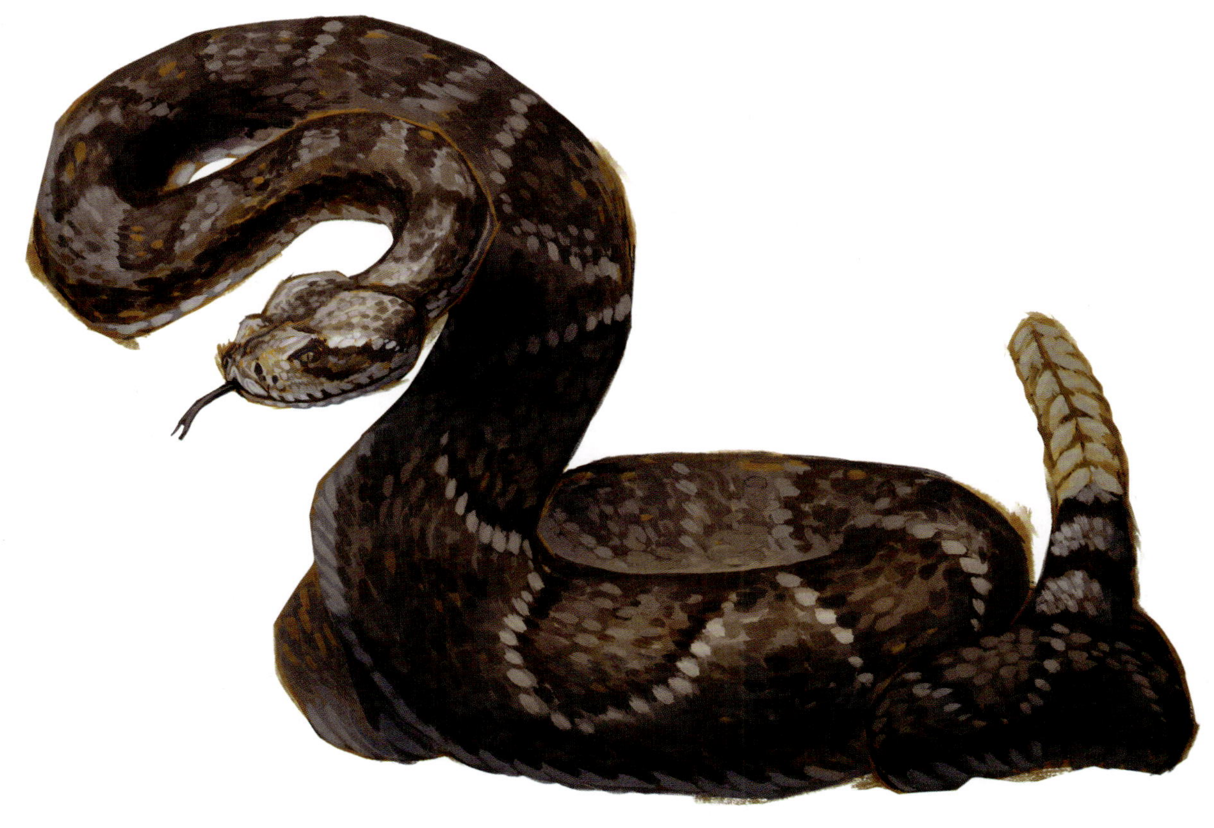

RATTLESNAKES

Location: Under rocks, on your driveway—anywhere and everywhere.

Considerations: Keep your wits about you. One bite from these beauties, and you're in big trouble.

Opposite: Mojave green—A beautiful pale-green snake with one of the world's most deadly neurotoxic/hemotoxic venoms. This snake can be found at the bases of Joshua trees in the Mojave, where its name hails from.

Above: Diamondback—The snake you're most likely to find dead and stuffed in a gift shop. The diamondback is the most famous of the rattlesnakes, and its bite is mostly responsible for all rattlesnake deaths in the Southwest.

Overleaf: Sidewinder—A deadly snake with a peculiar way of moving. This snake can be found coiled around bushes, out on the sands at night and along railroad tracks.

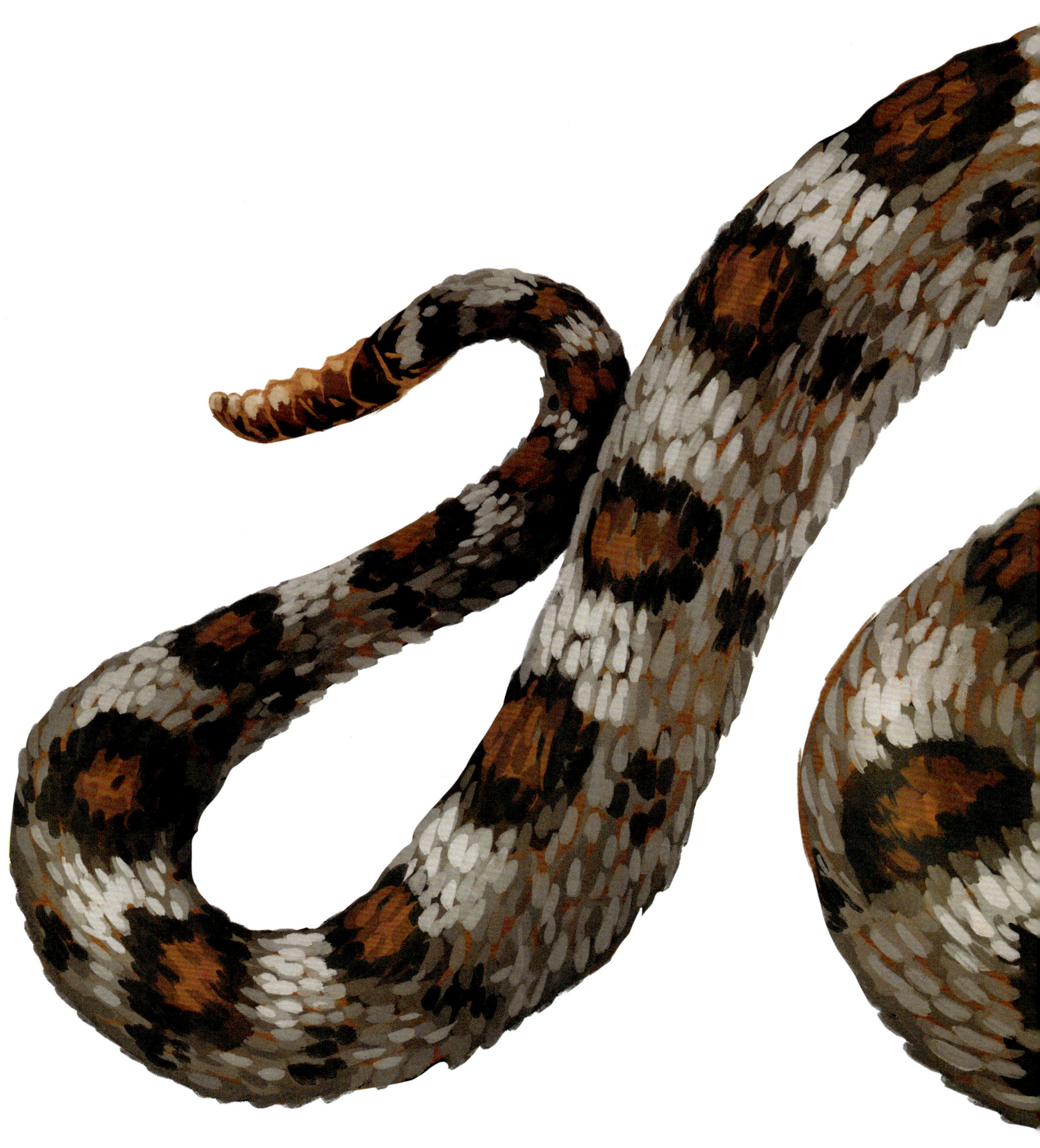

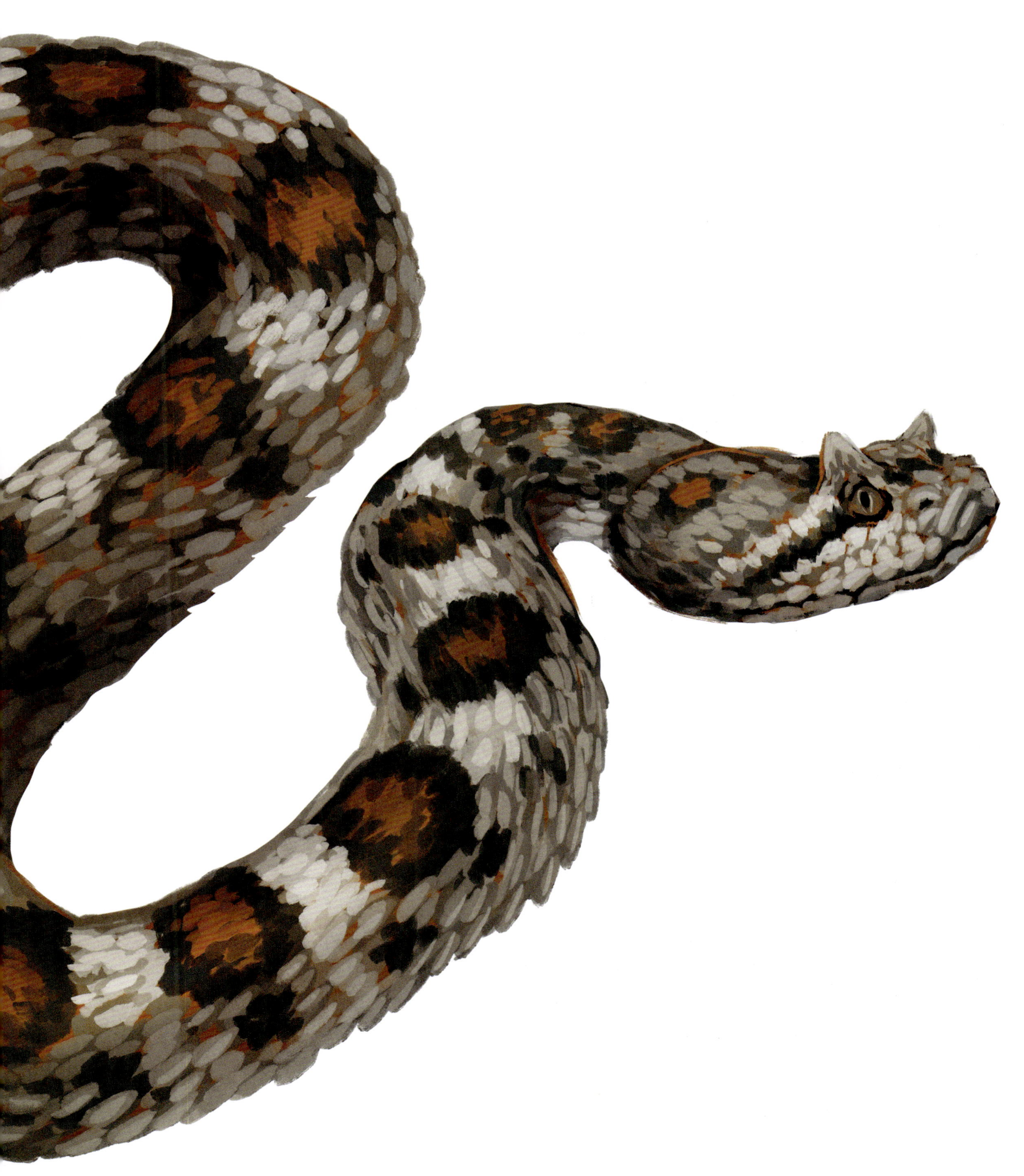

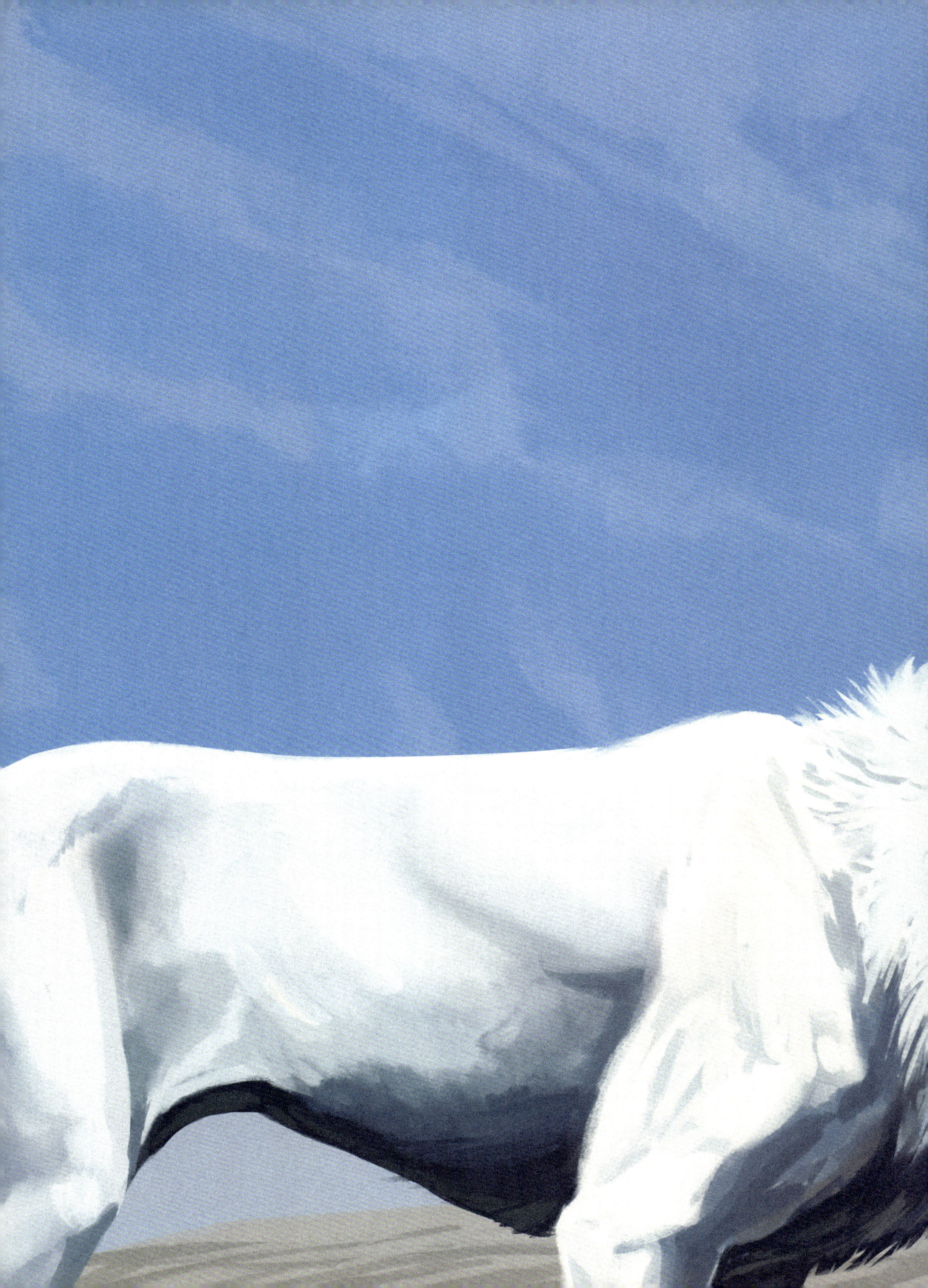

WHITE LION

Location: 34.5602, -115.6423

Considerations: Bring an offering for this one.

Legend has it there's a white lion that roams the desert around old Route 66 in California. It could be an escaped movie star from Hollywood. It could be a mirage. It could be that marble statue near highway 66 that some lost traveler saw in the heat. Dehydration is one hell of a muse.

Either way, if you're headed out that way, bring a gift.

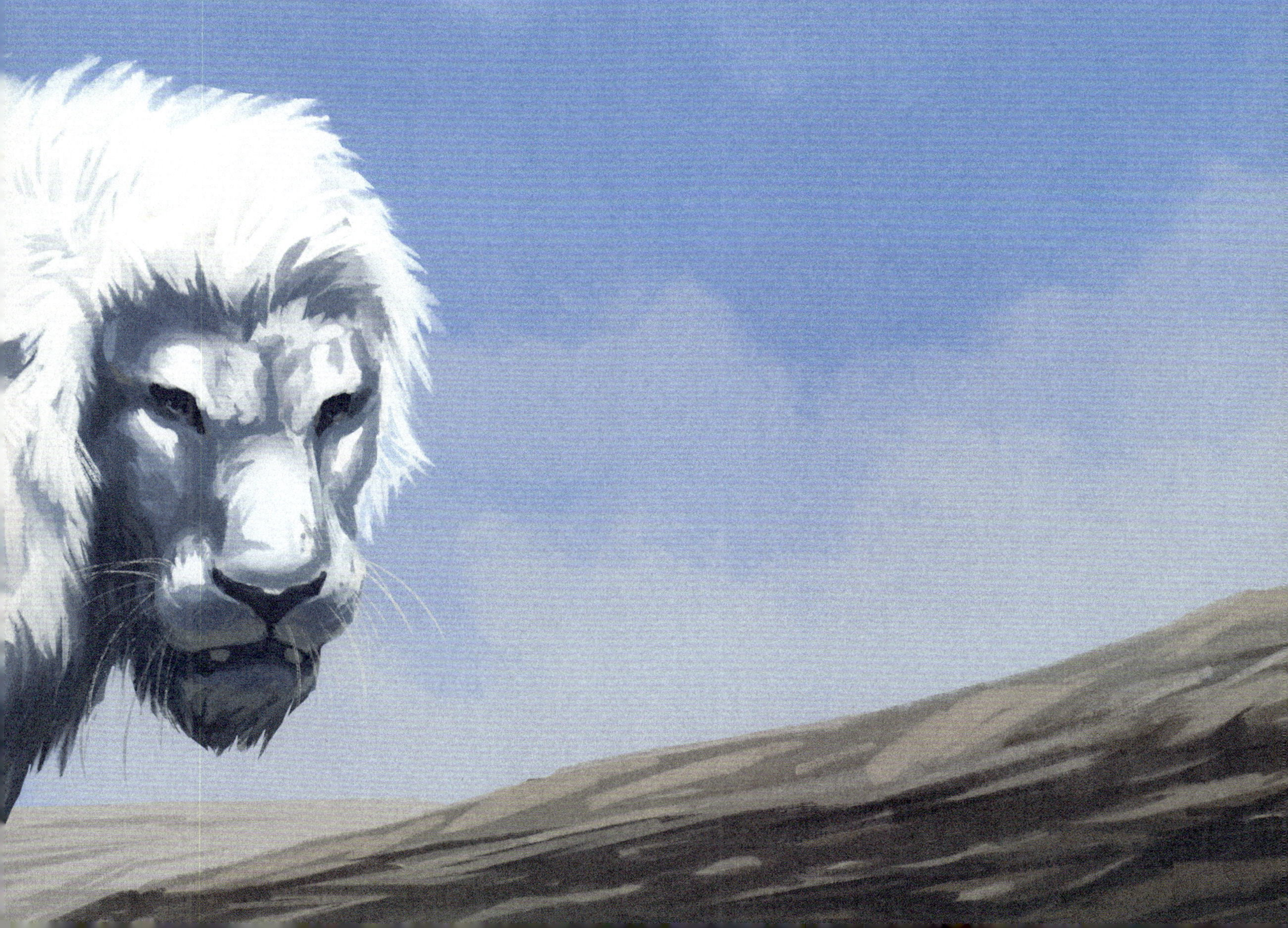

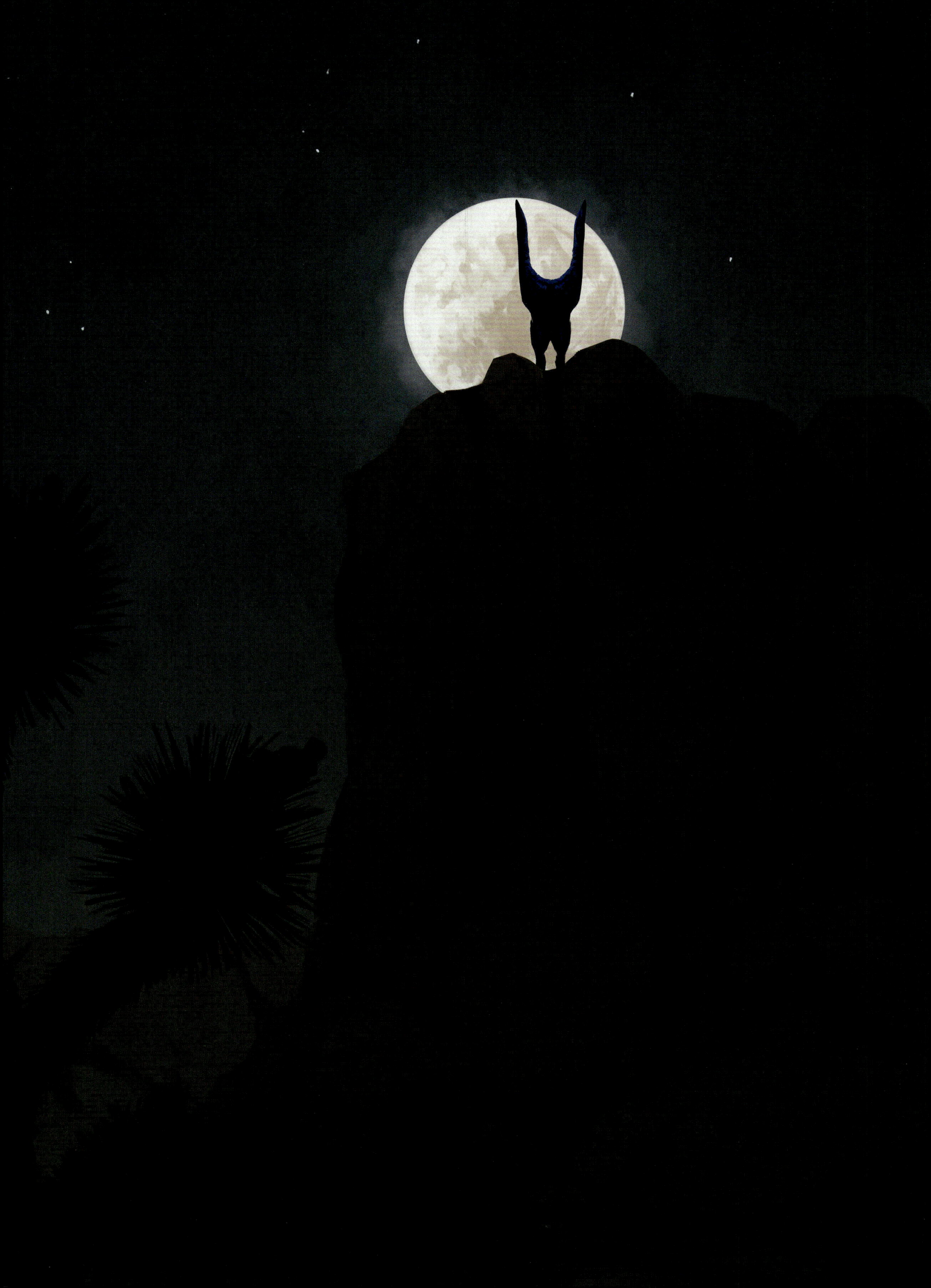

PRONGS

Location: Twentynine Palms.

Considerations: Make note of their position and bring a garbage bag.

Some call them Prongs—odd creatures that live among the rocks and boulders near the southernmost edge of the Mojave Desert, out near Twentynine Palms. Prongs have two legs and a large split head that fans out like a fork. They are estimated to stand around six feet tall and have been described as a deep indigo blue. They are usually seen from a distance, standing silently among the rocks and staring into the desert.

They are often seen in the light of a full moon, when the shadows cast across the craggy boulders like ink across a page. To spot one, keep your eyes on a group of shadows and wait until the shadows shift as they adjust their stance. In the still night, their prong-headed silhouette can be spotted among the surfaces of the formations. Make note of their position and do not approach them. Some people have reported an uncomfortable, loud vibration that emanates from them if they are engaged.

Prongs didn't appear in the deserts of Twentynine Palms until the heavy development did. Whatever the case, these creatures started showing up around the 1950s, during the construction of the Marine Corps Air Ground Combat Center (MCAGCC). Disturbances from this settlement—along with the surrounding housing tracts and strip malls—upset something, and now these strange beasts appear occasionally. Some think these creatures are an ecological reaction to the landscape's trash, toxic metals and garbage. Others think that the creatures thrive on it.

There *is* a lot of garbage in the desert. If you've done any sort of looking around, you certainly have come across things like shot-up soda cans, old mattresses and broken refrigerators baking along the road. People typically see desert land as open, empty space, devoid of life and value—a prime place for dumping garbage, broken appliances and maybe things you don't want to remember. The prongs remember.

At sunrise, piles of rusted metal objects can sometimes be found where the prongs were standing. It's not clear why they collect such objects or leave them behind, but it is advised that the piles are only removed with caution.

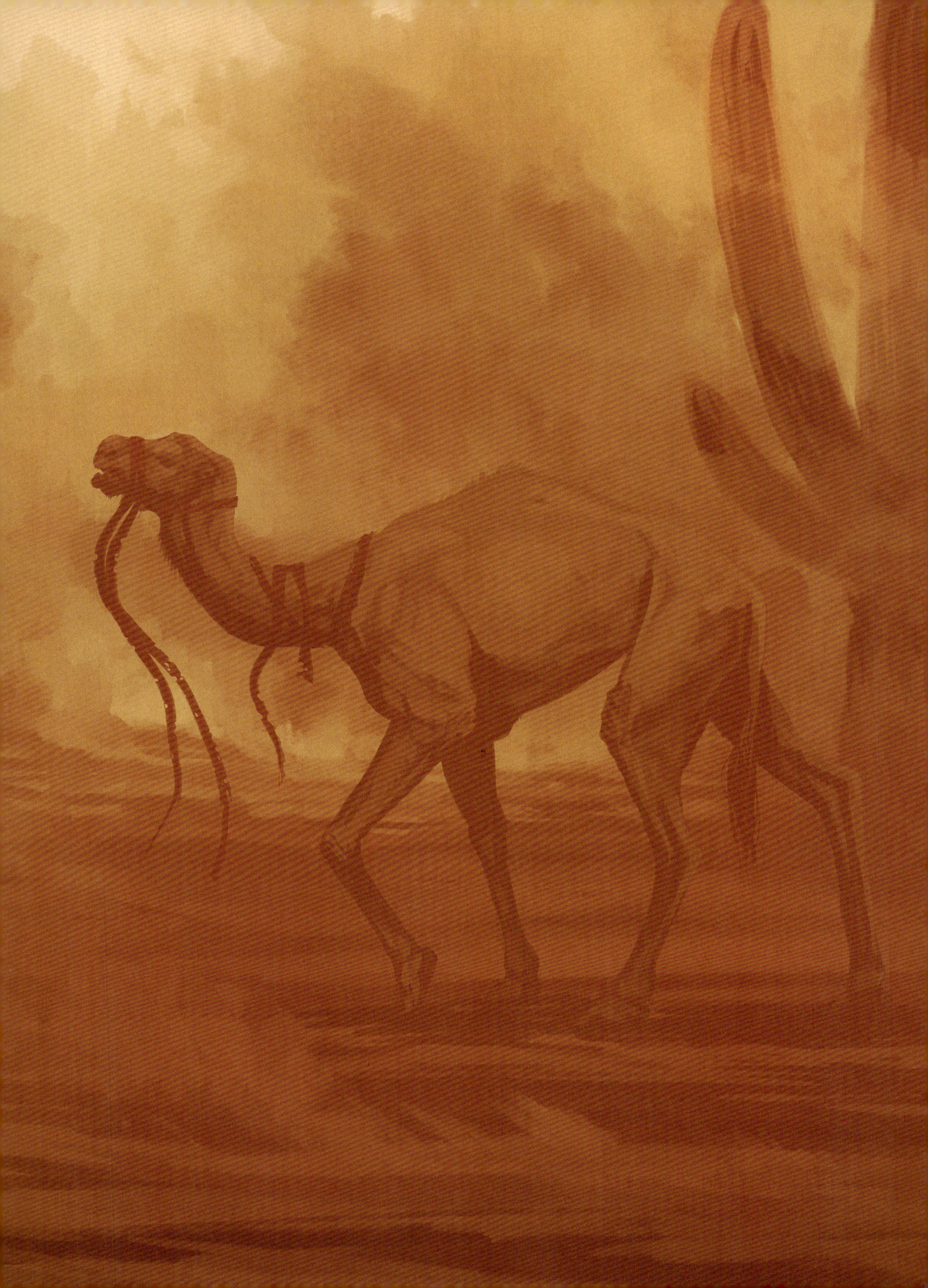

WILD CAMELS

Location: Omitted.

Considerations: Seeing this one is a curse.

There are wild camels that roam the open expanses of Nevada and Arizona. They carry with them ghosts of railroads.

Camels haven't been present on the North American continent for eons and yet they found themselves transported by the US army to Texas where they would be used as beasts of burden.

The camels that have been spotted are tangled in leather lashings, leftover from an unknown rider. They are not to be approached as they can be extremely dangerous.

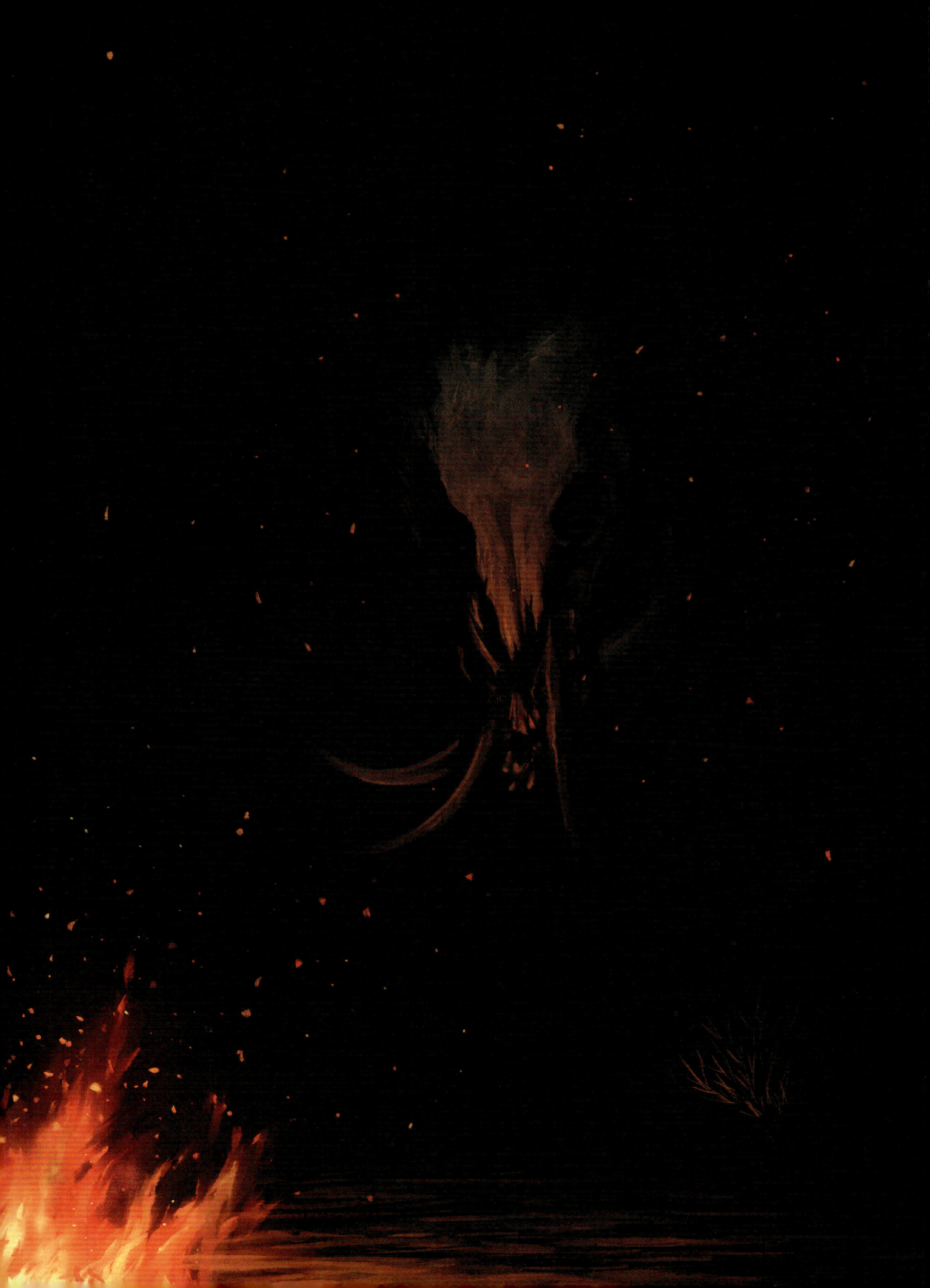

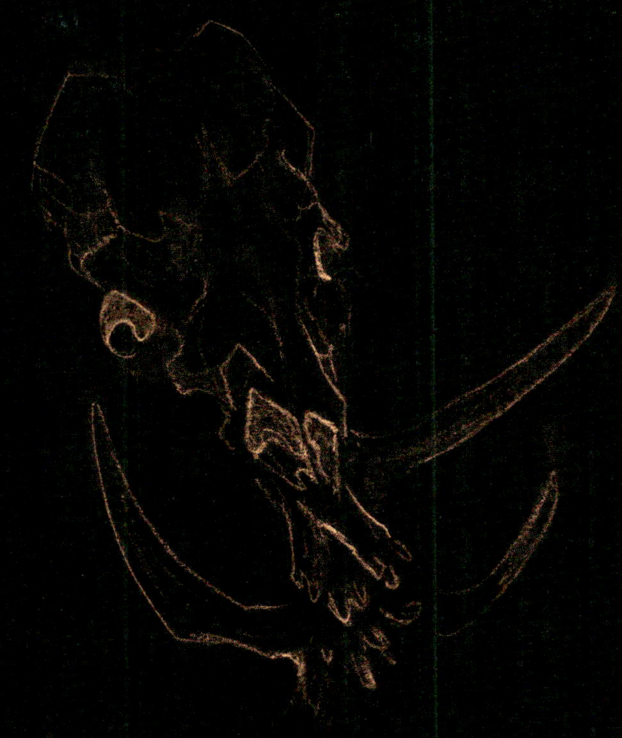

FIRE SPIT

Location: Desert backcountry

Considerations: If you see a spit, look at the moon. She will cleanse your eyes. If there is no moon, you're out of luck.

Beware the campfire in desert backcountry. Letting your guard down can invite a spit, a grotesque and invasive specter. Spits take on horrifying and distorted forms as they court the campfire light.

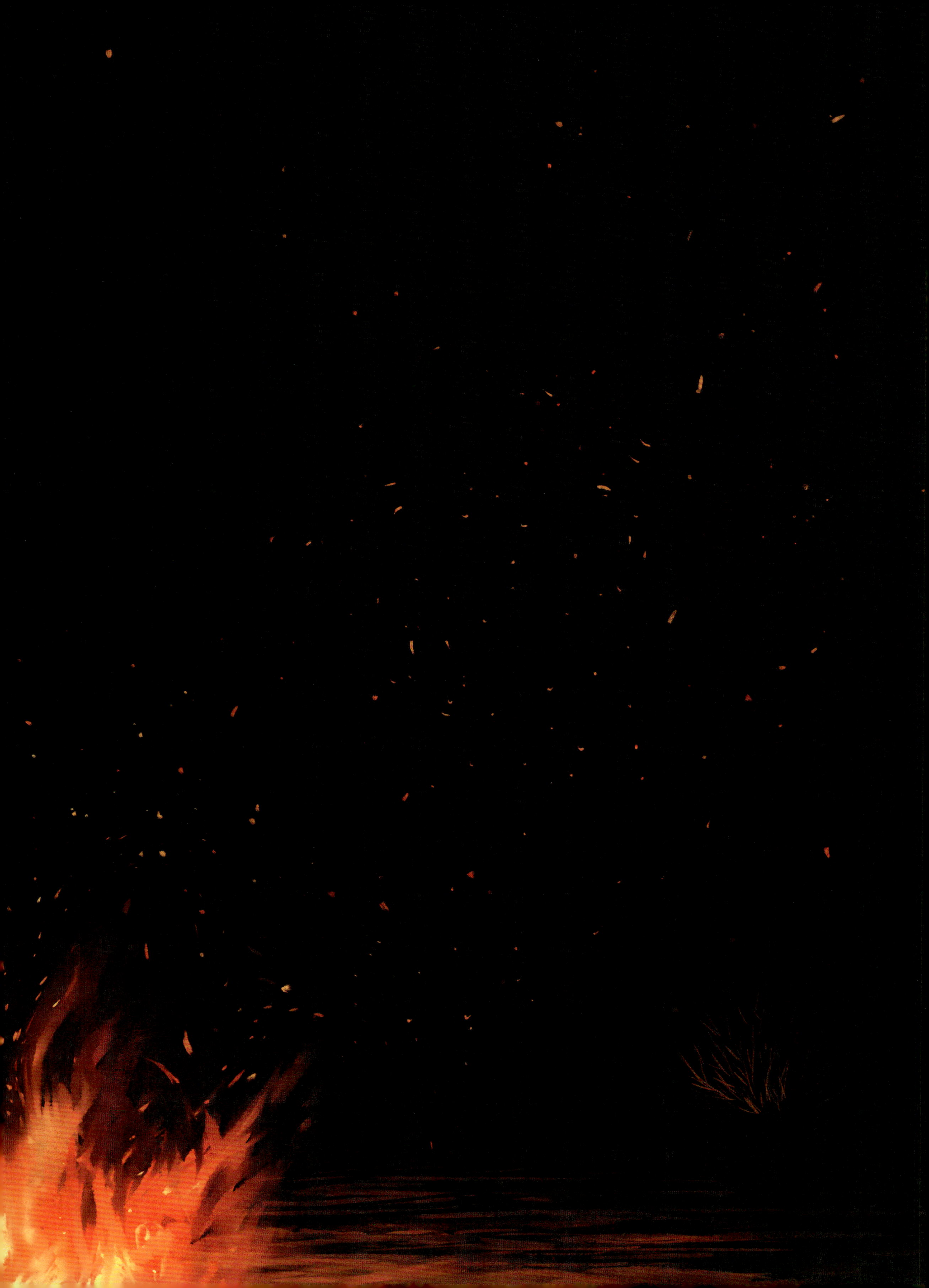

THEY SIT JUST AT THE EDGE OF YOUR VISION, AND THEY'RE ALREADY GONE BY THE TIME YOU'VE BEEN SPOOKED.

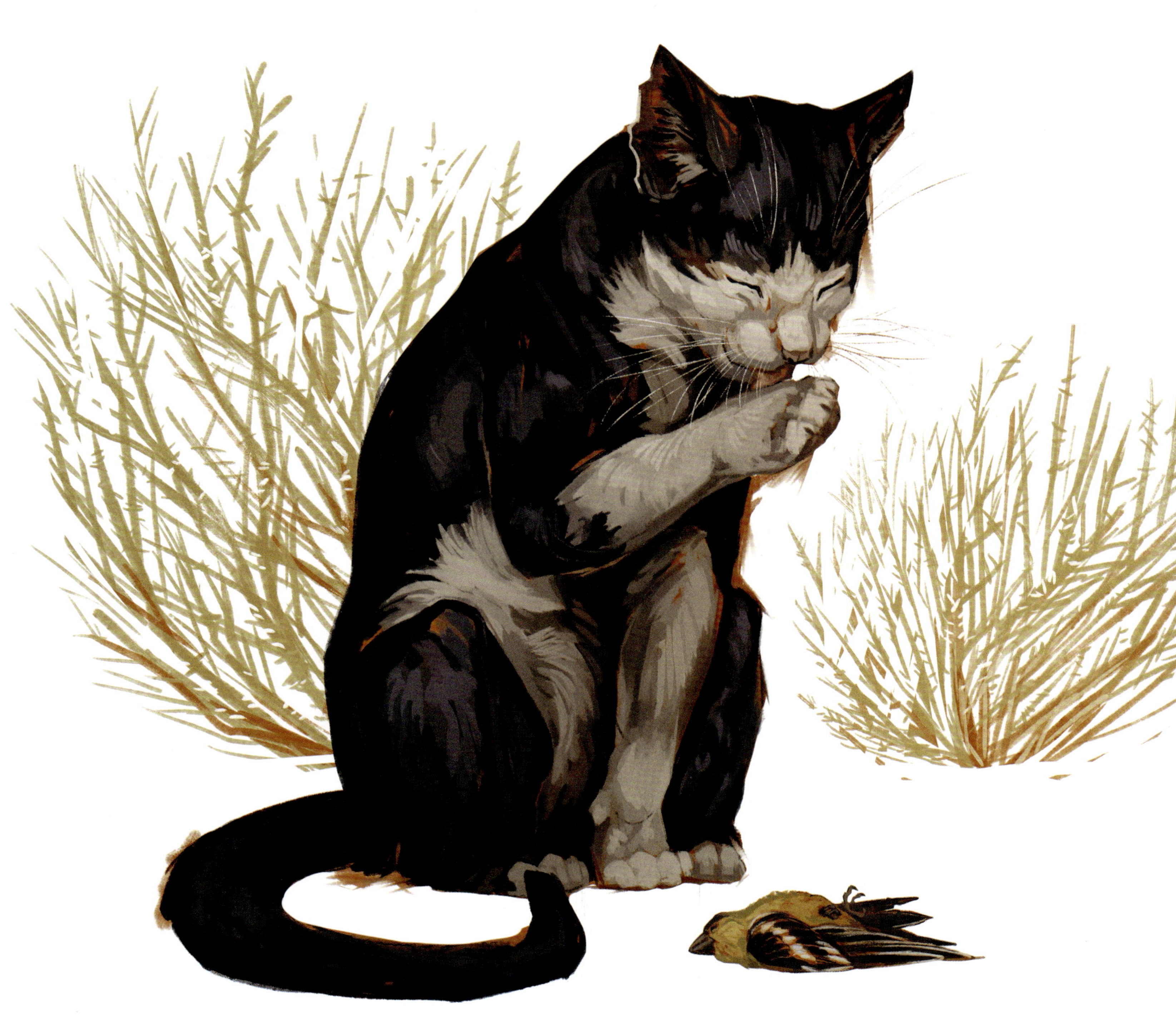

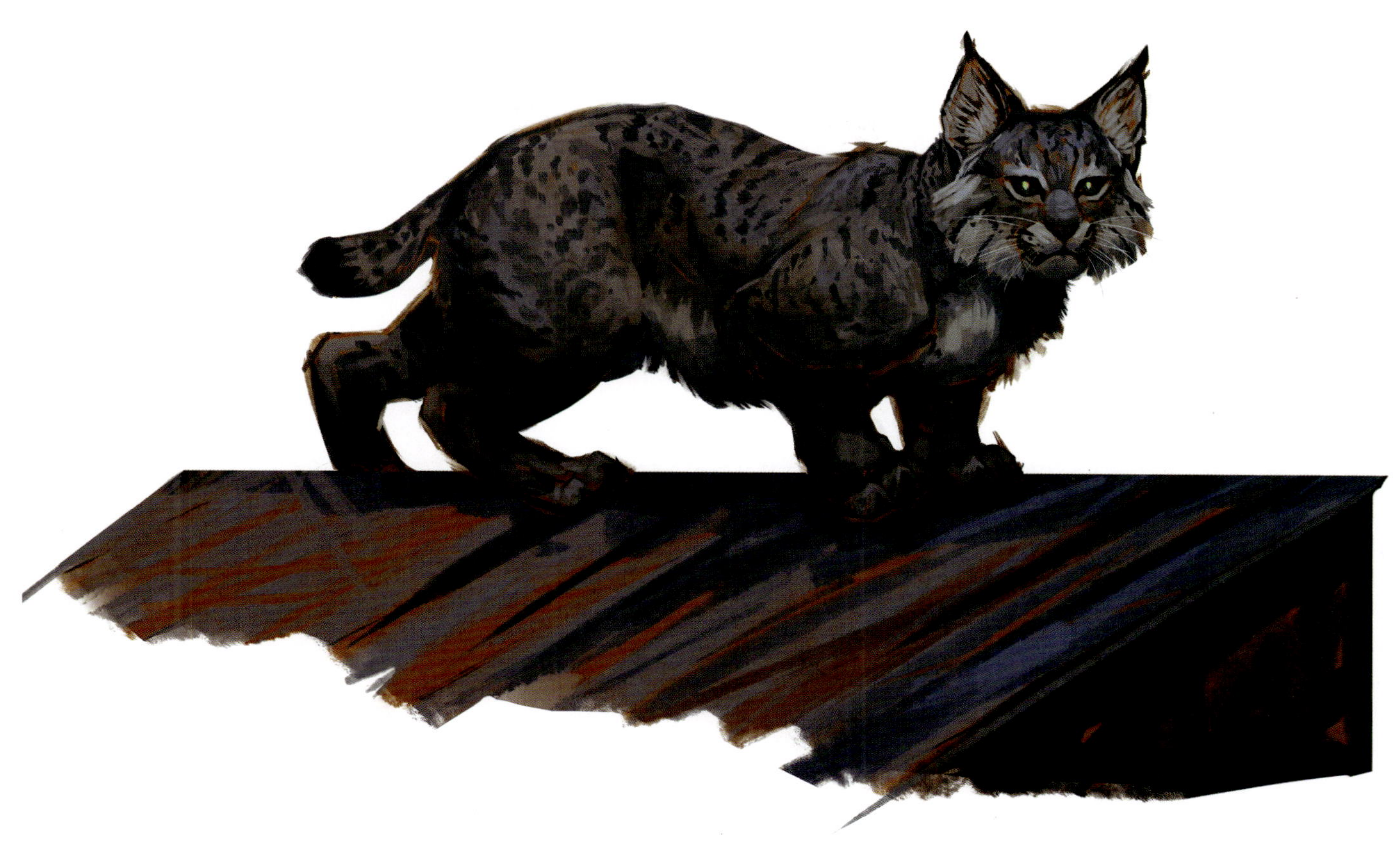

DESERT CATS

Locations: California, Utah, Nevada, Arizona, New Mexico and beyond.

Considerations: Desert cats are dangerous, even the little ones.

There are seven cats that inhabit the southwestern deserts of North America.

The first is the house cat (*opposite*). Feral, stray or pet, it is the most successful of the cats.

The second is the bobcat (*above*). A small but fierce predator, it preys on small animals and neighborhood chicken coops in the dim light of morning.

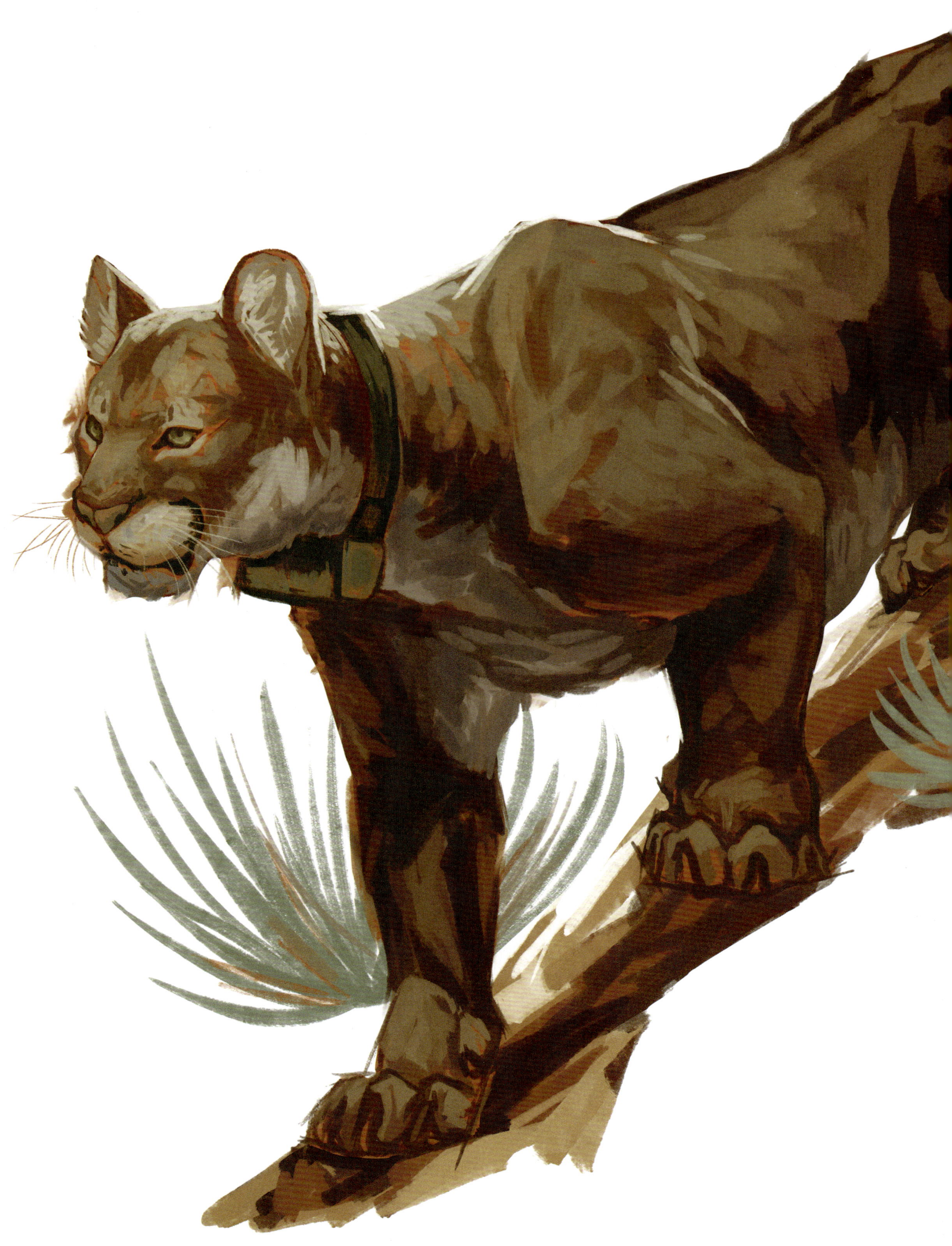

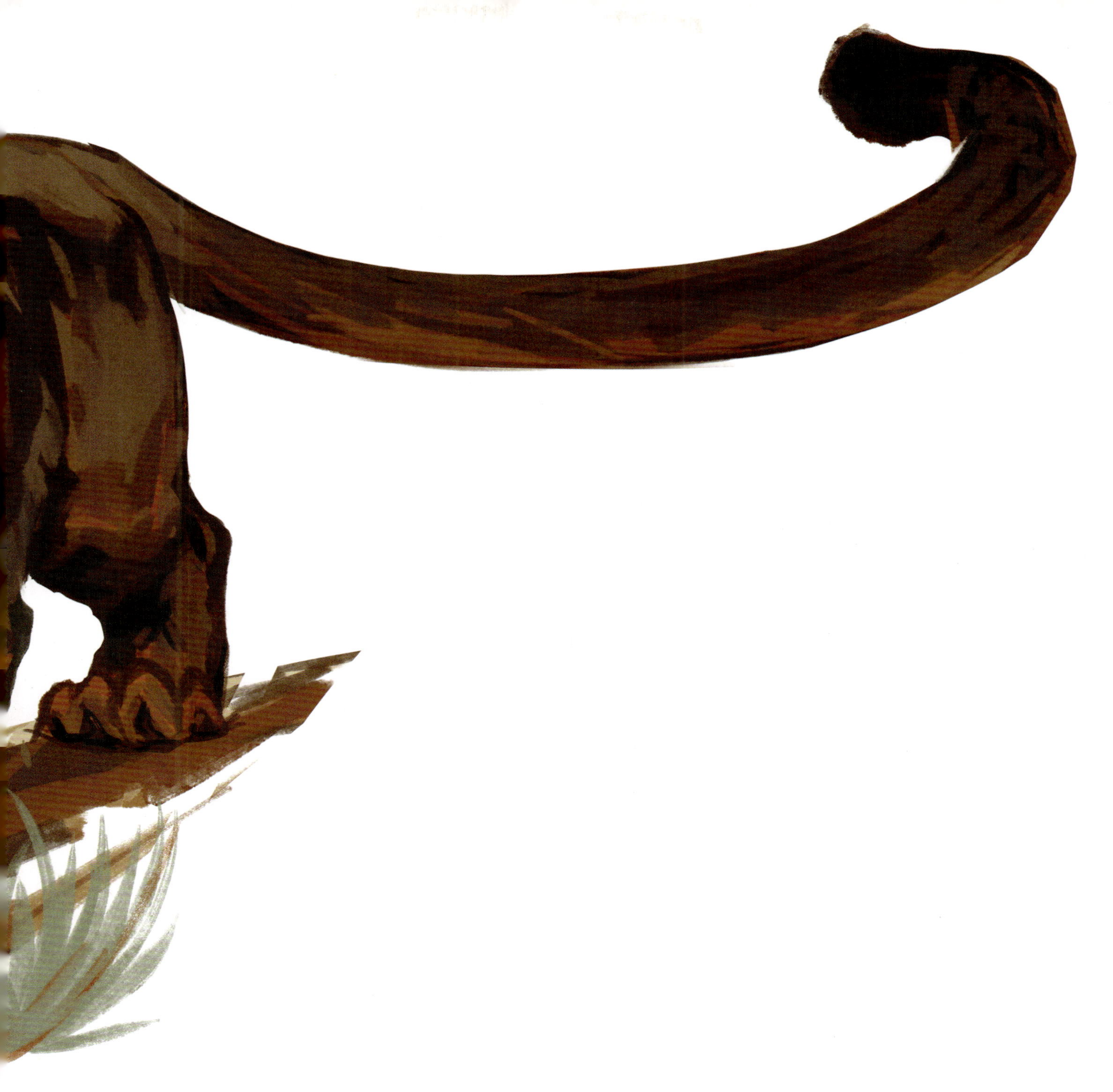

THE THIRD IS THE MOUNTAIN LION. PUSHED INTO SMALL CORNERS OF WILDERNESS, THIS LARGE PREDATOR SLINKS ABOUT IN REMOTE FOOTHILLS AND THE IMAGINATIONS OF HIKERS.

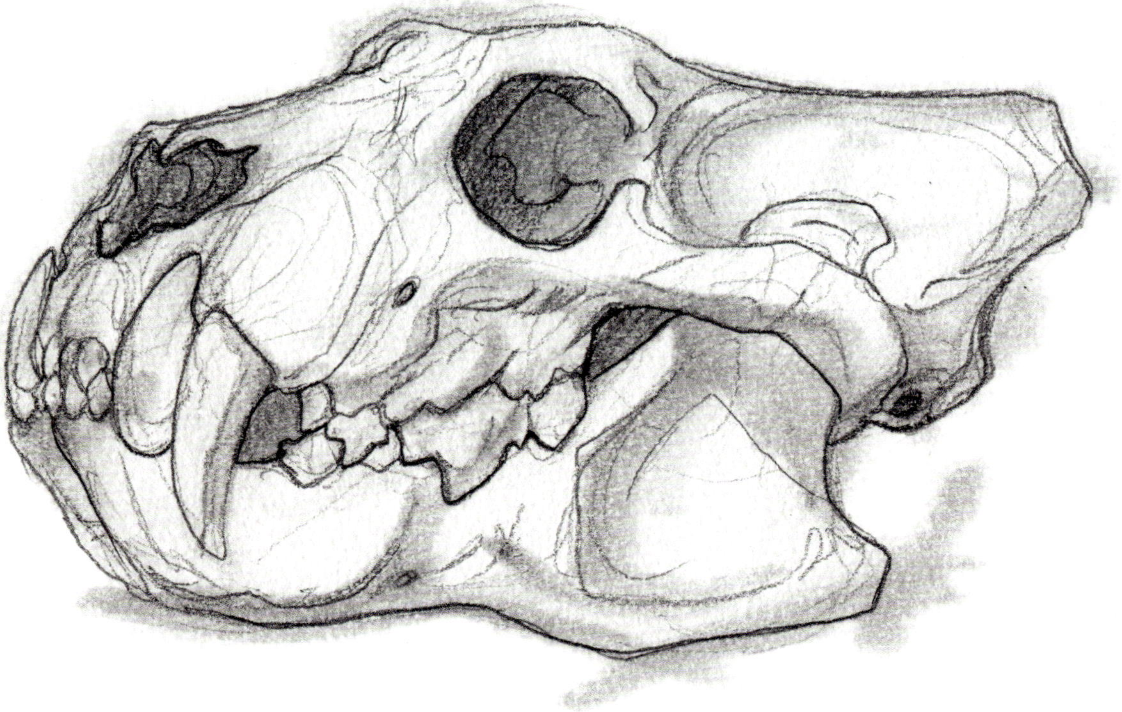

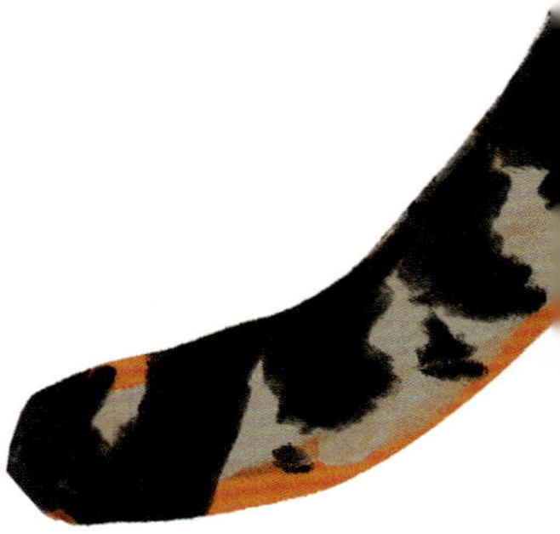

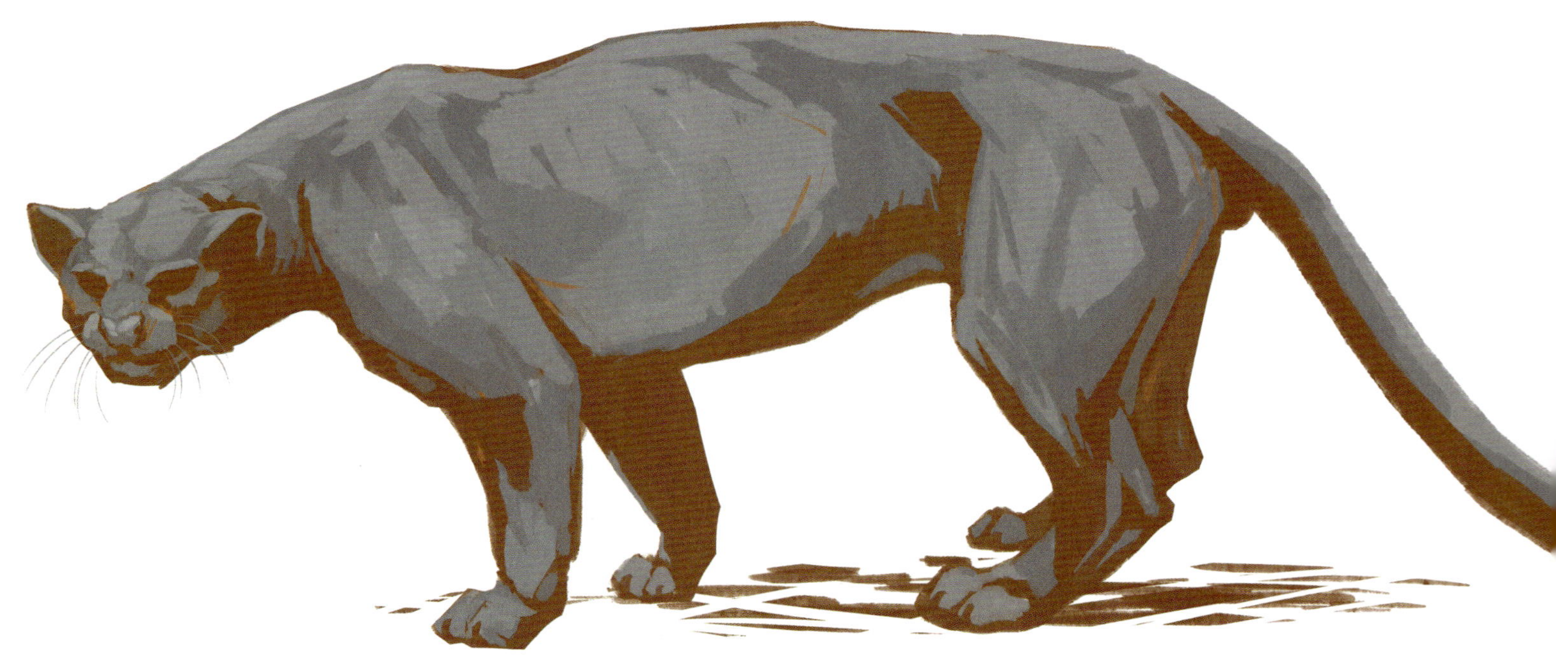

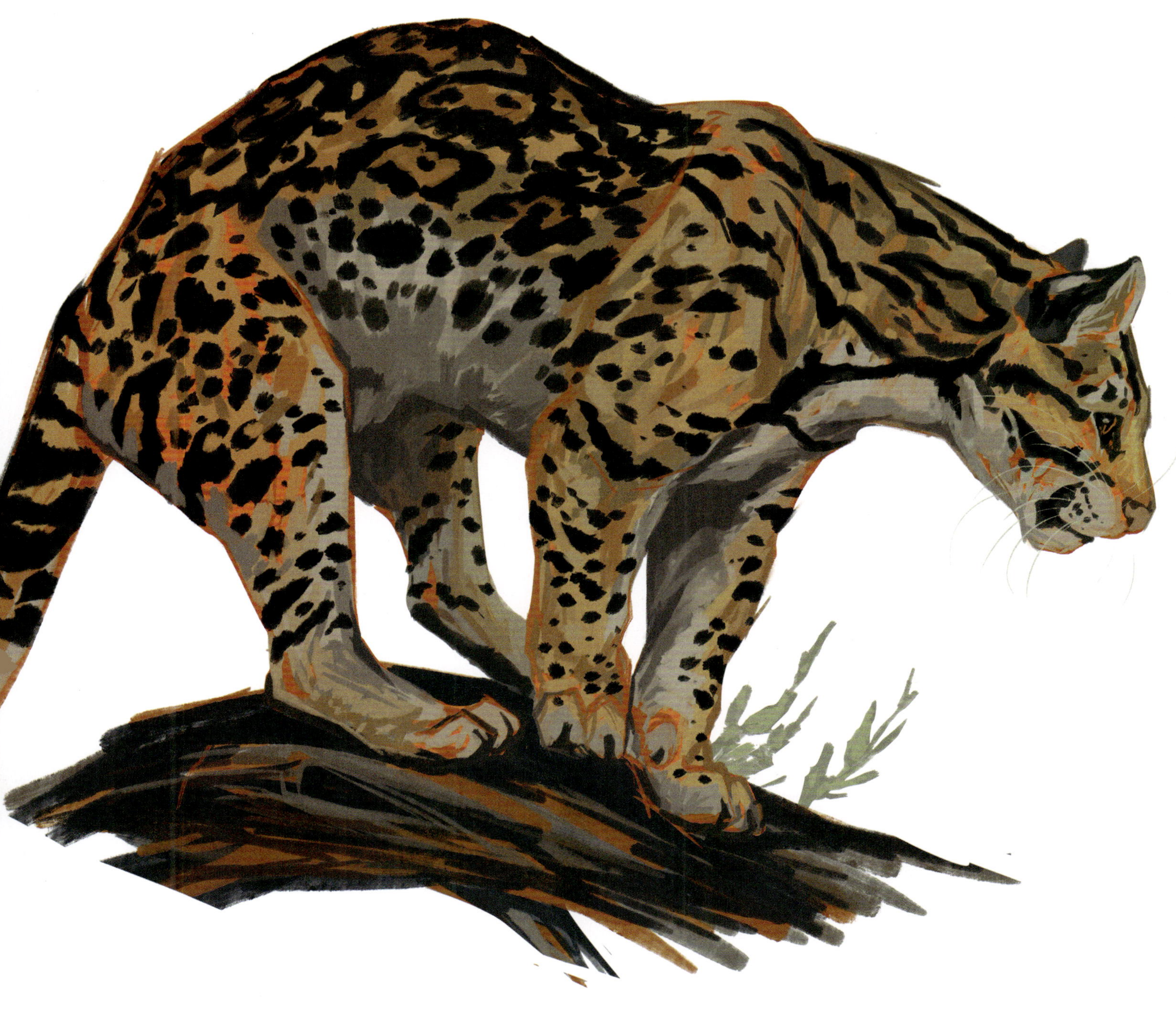

The fourth is the jaguar (*opposite top*). There have been a few sightings of individuals that may still live in Arizona but otherwise it has been extirpated from its habitat in the U.S.

The fifth is the jaguarundi (*opposite bottom*). An ancient looking animal that's all but a ghost in the Southwest these days.

The sixth is the ocelot (*above*). A small, spotted cat with a prized pelt, this poor beast is spread thin over its broken habitat and is sold illegally in the pet trade.

THE SEVENTH CAT IS THE ONE YOU'D RATHER
NOT ENCOUNTER.

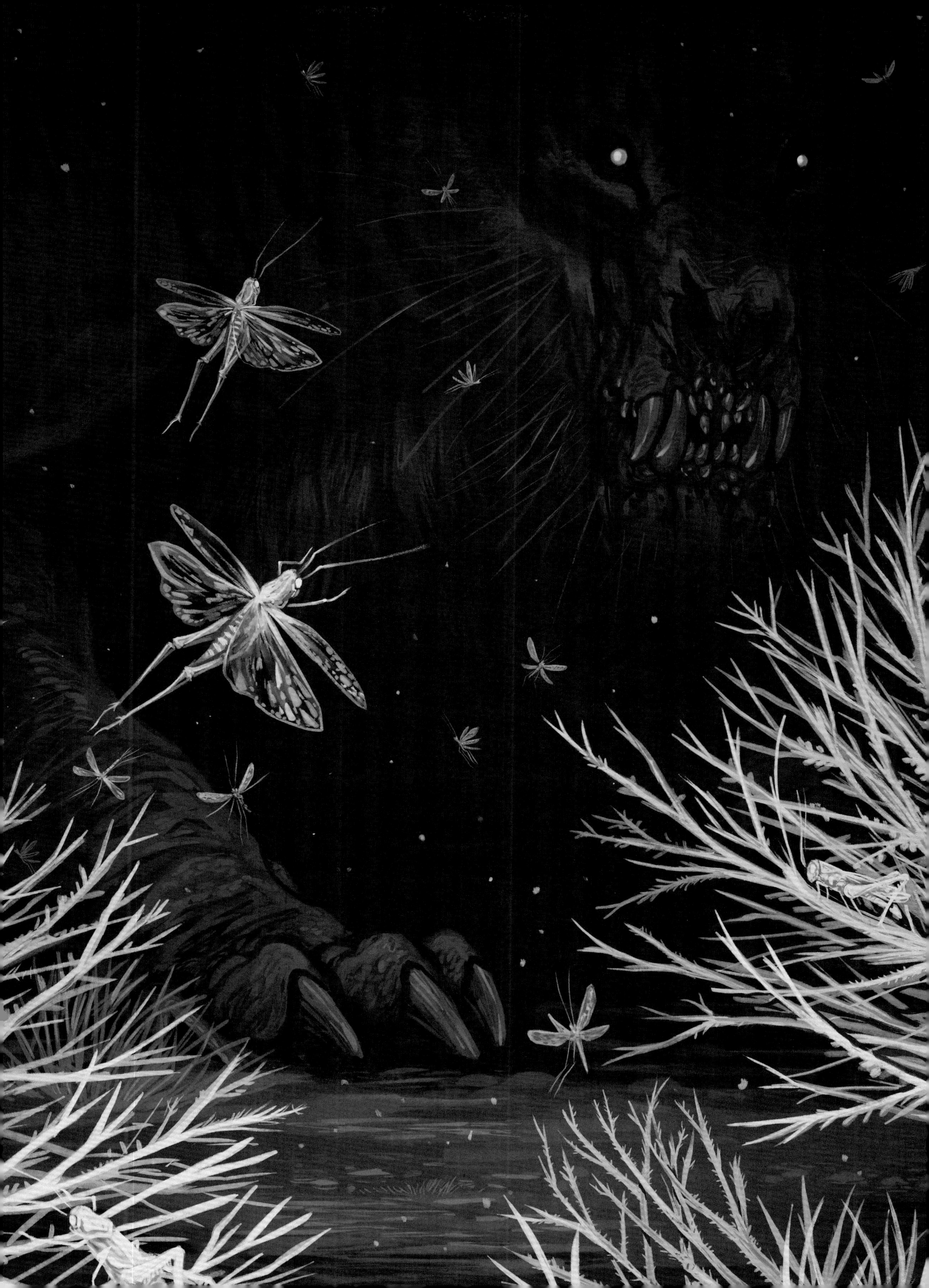

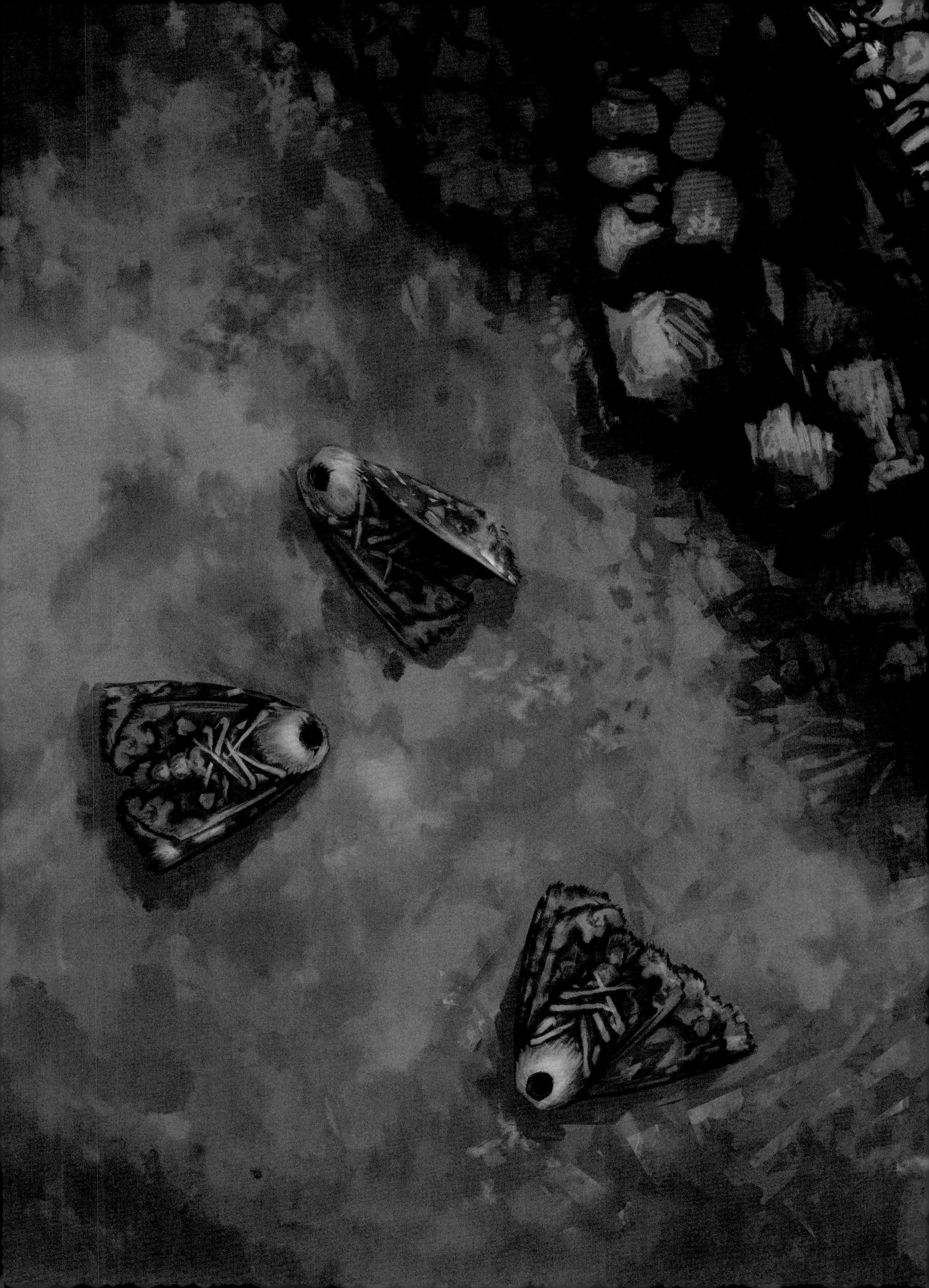

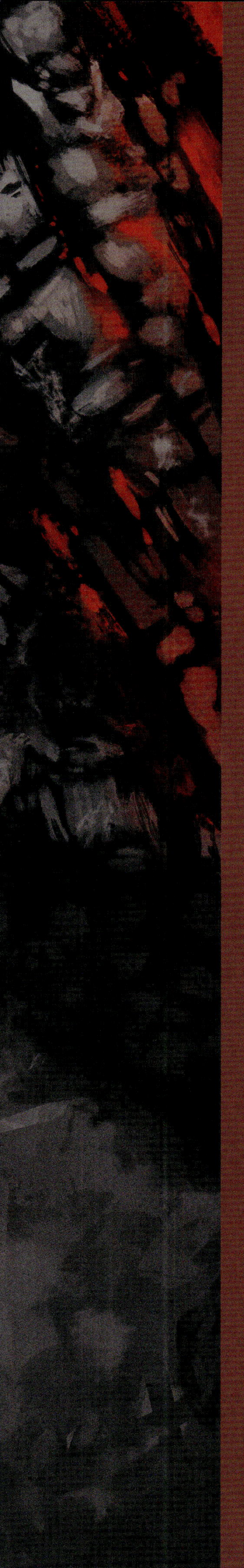

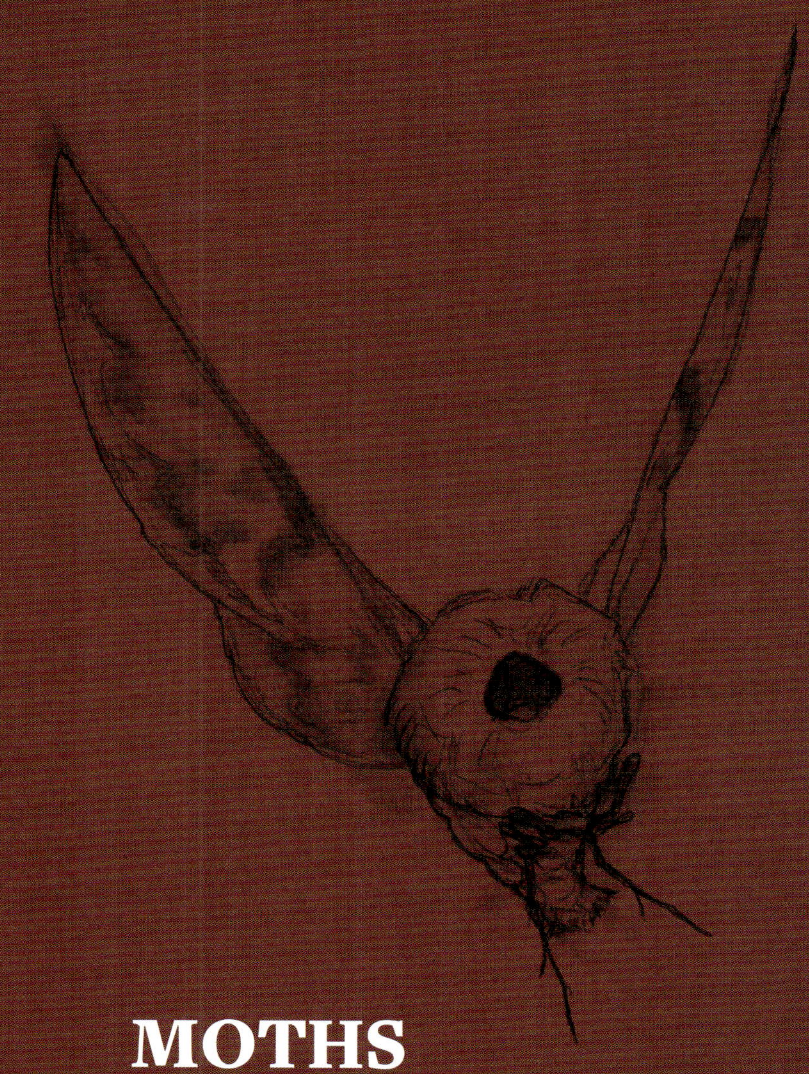

MOTHS

Location: Northern Arizona.

Considerations: None.

Like small paper pieces, these little creatures flutter around the edges of my campfire. They resemble moths, with angular wings and tufted bodies.
In the morning, I find a few dead ones next to the embers.
No heads.

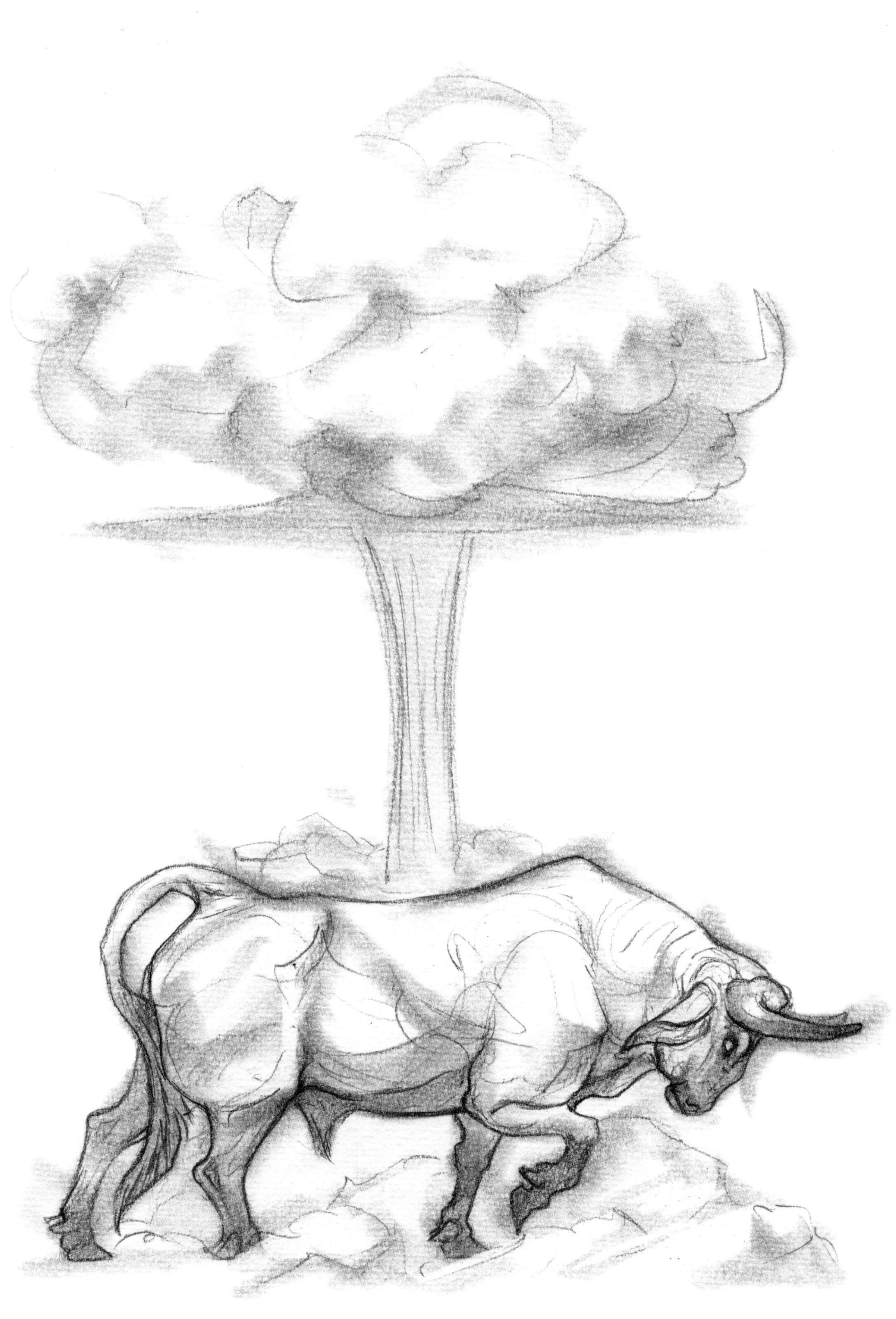

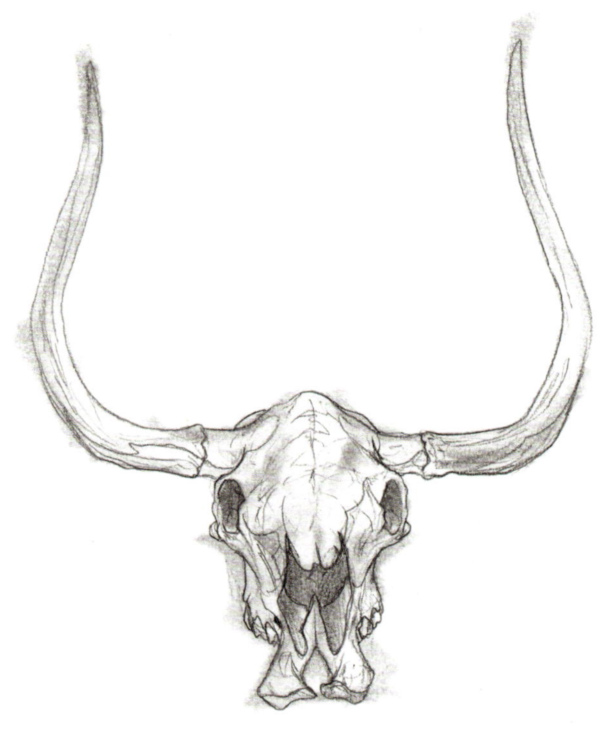

THE BLACK STEER

Location: Alamogordo Bombing Range, known as the *Jornada del Muerto*.

Considerations: There are tours and plaques and such, but I'd recommend avoiding the place.

The Manhattan Project conducted its first nuclear test in New Mexico, just 210 miles south of Los Alamos. Upon detonation, the plutonium implosion device released 18.6 kilotons of power, and the surrounding sands were instantly turned into green glass.

Ranchers in the area reported bleeding and burns on livestock from the fallout. The local population suffered cancers and other illnesses and deaths months and years after the explosion, which they were never warned about.

Since then, old curses walk the land, bearing witness to the way things changed.

"Horns shiny-black and eyes blood-red, the black steer cometh when we're all dead."

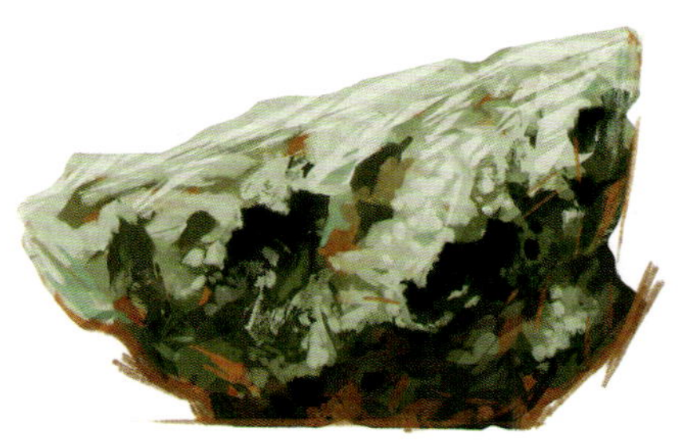

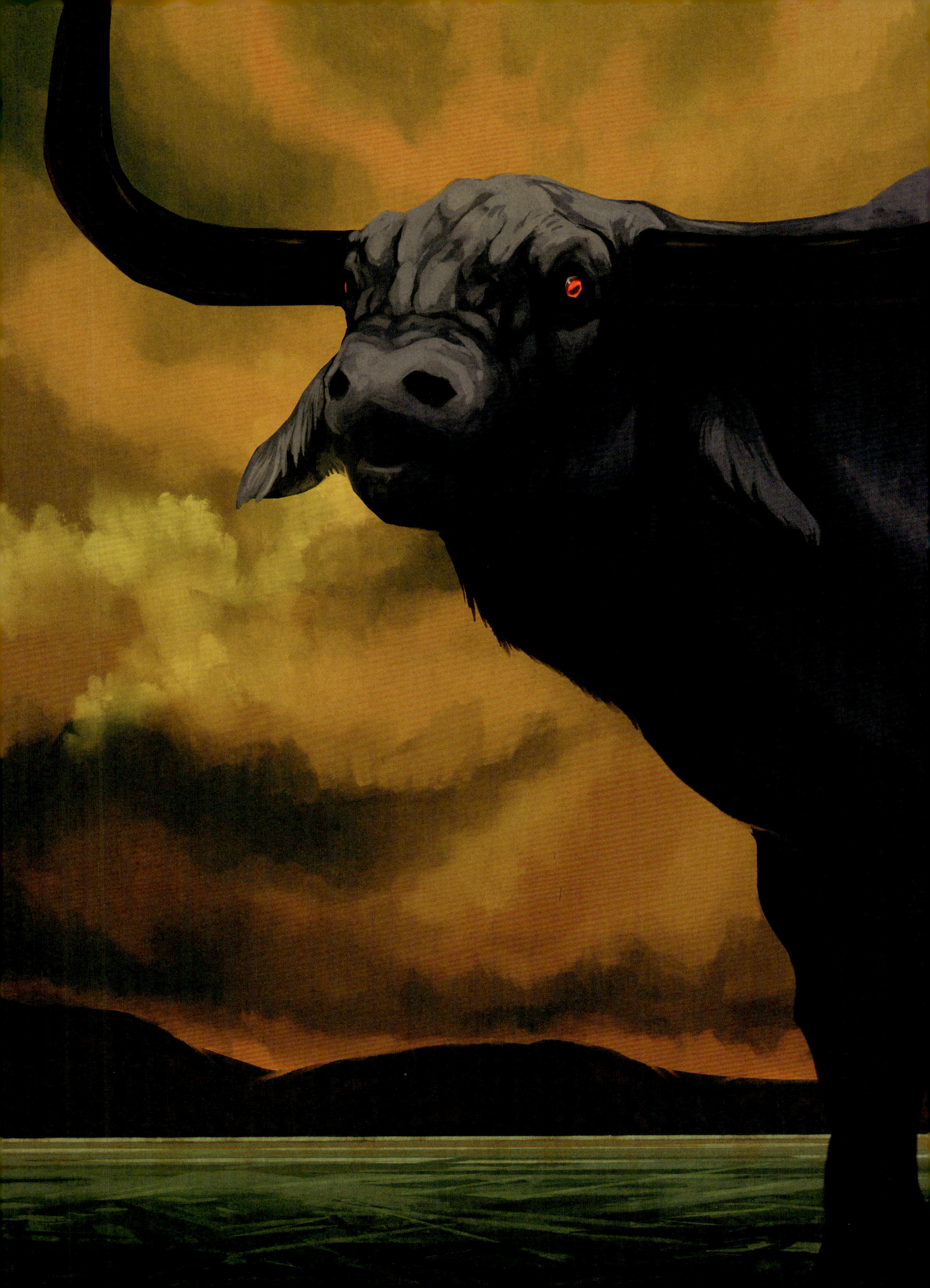

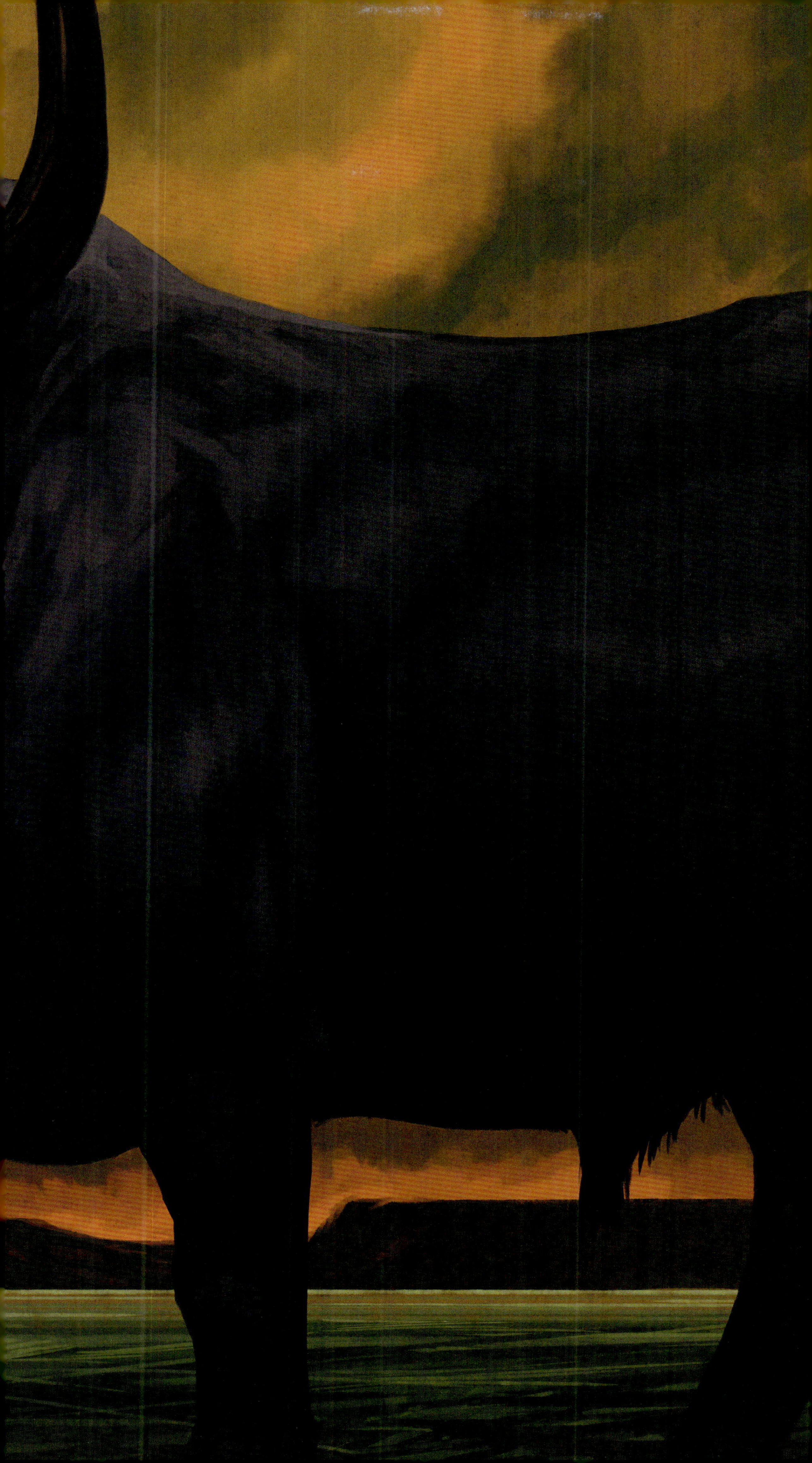

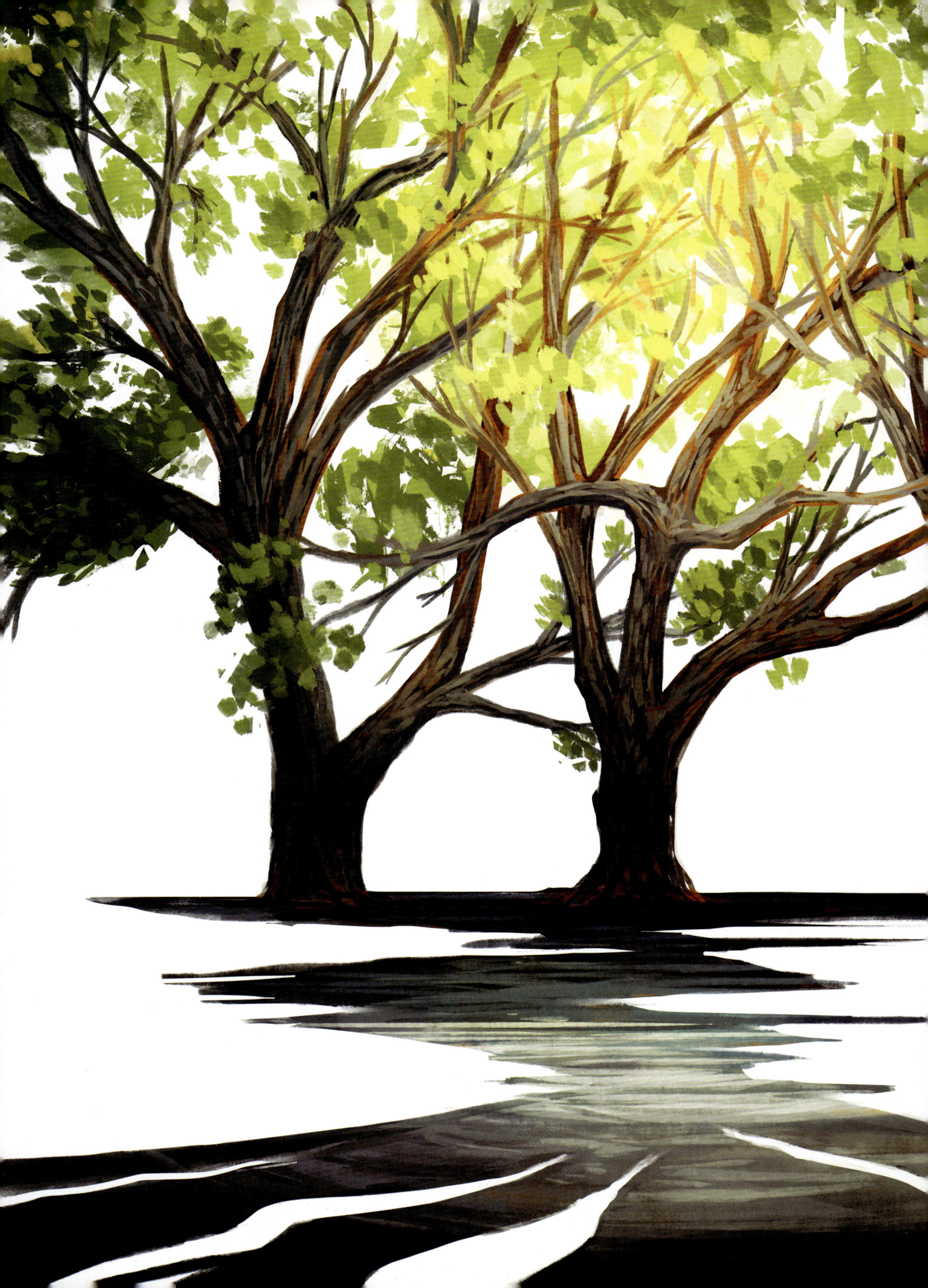

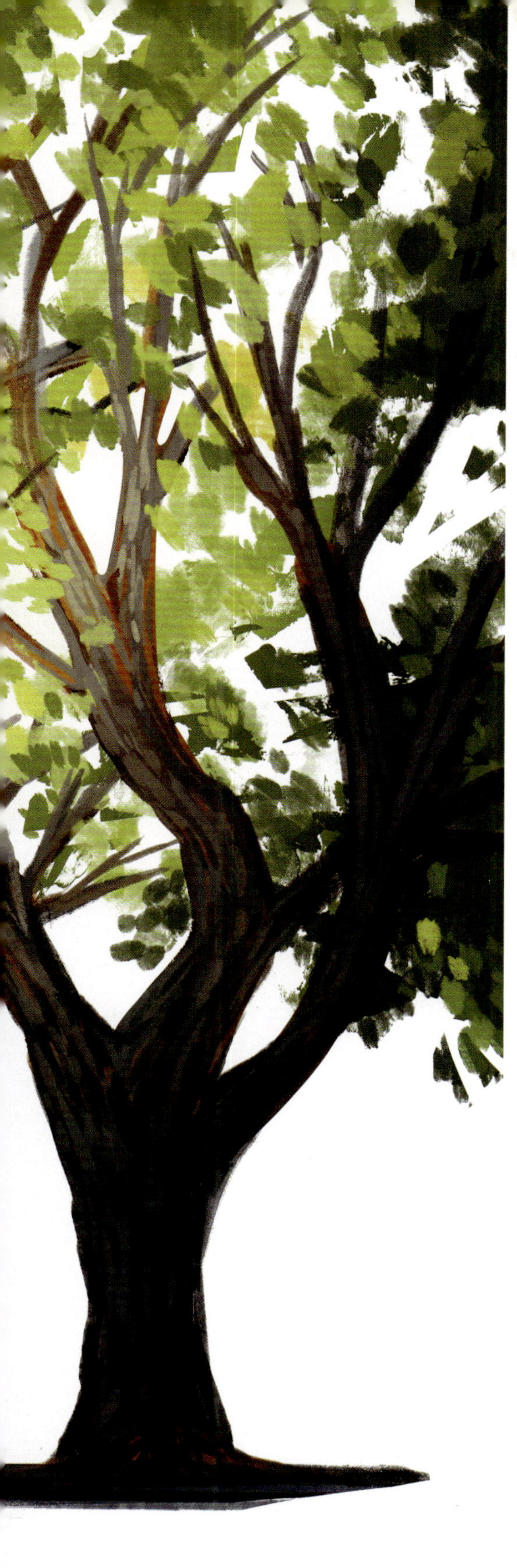

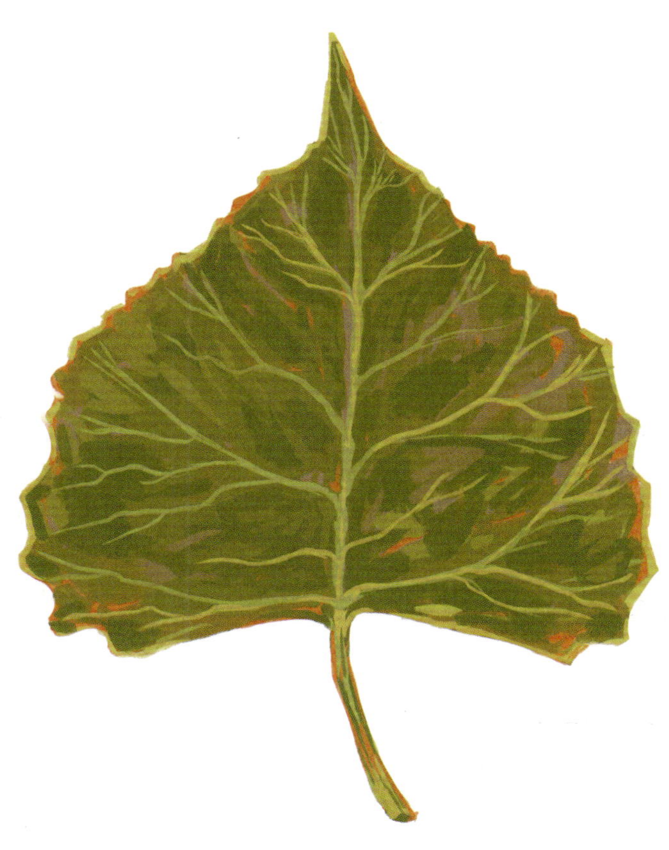

COTTONWOOD TREES

Location: Throughout the Southwest.

Considerations: These trees have been an indication of water to many people traveling through the desert. They grow along arroyos and near springs. If you see a cottonwood, you're in luck.

In a dream, I saw three cottonwoods.
    A small, emerald river flowed from their roots.
    Fountains in the dust.

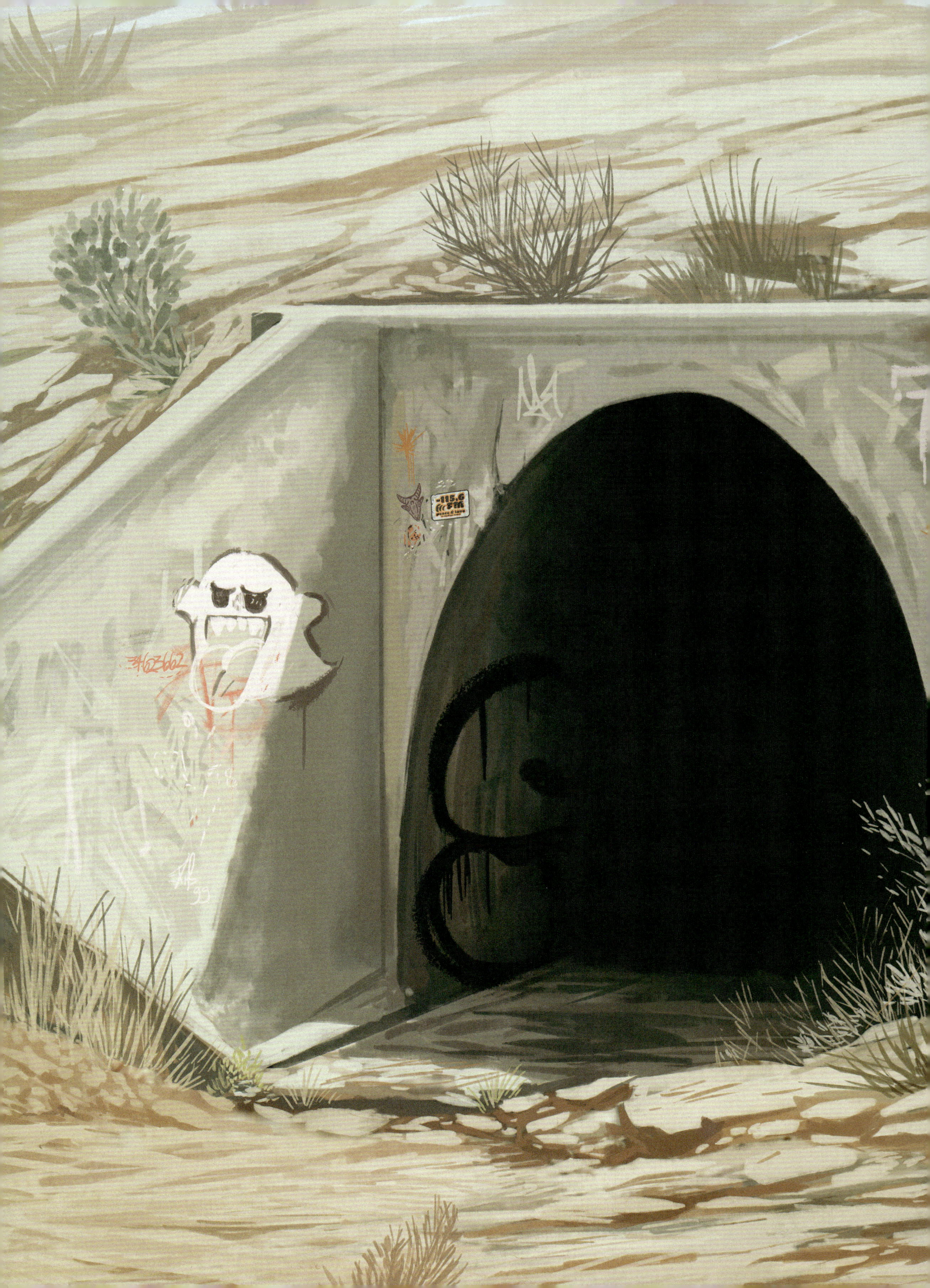

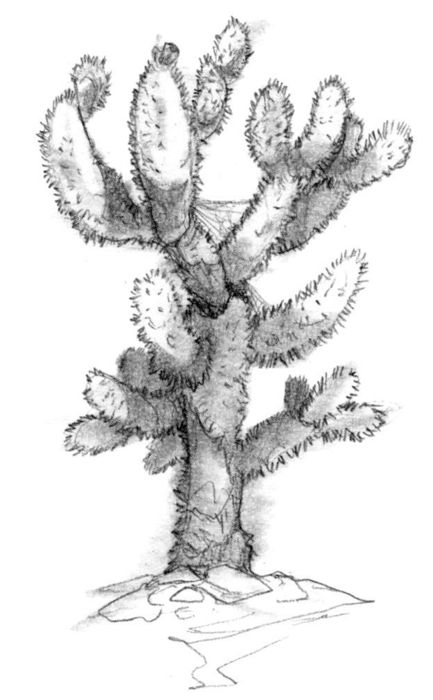

TUNNELS

Location: Many desert towns have a tunnel. The local kids will know where they are.

Considerations: Don't choose the wrong one.

Tunnels are tempting. A respite from the heat and perhaps a shortcut to somewhere. Keep your wits about you. Many have gone missing in these portals.

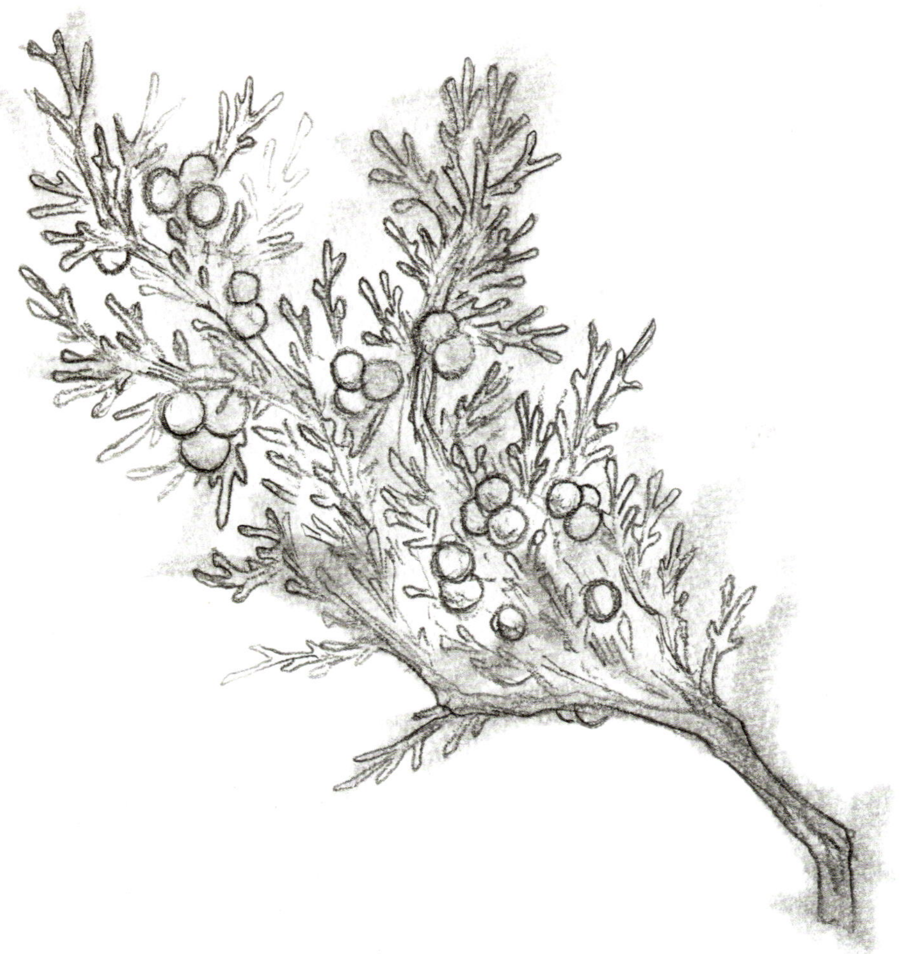

THE HIKER

Those wide blue skies
and that clean, scorched earth.
Did you love it, friend?

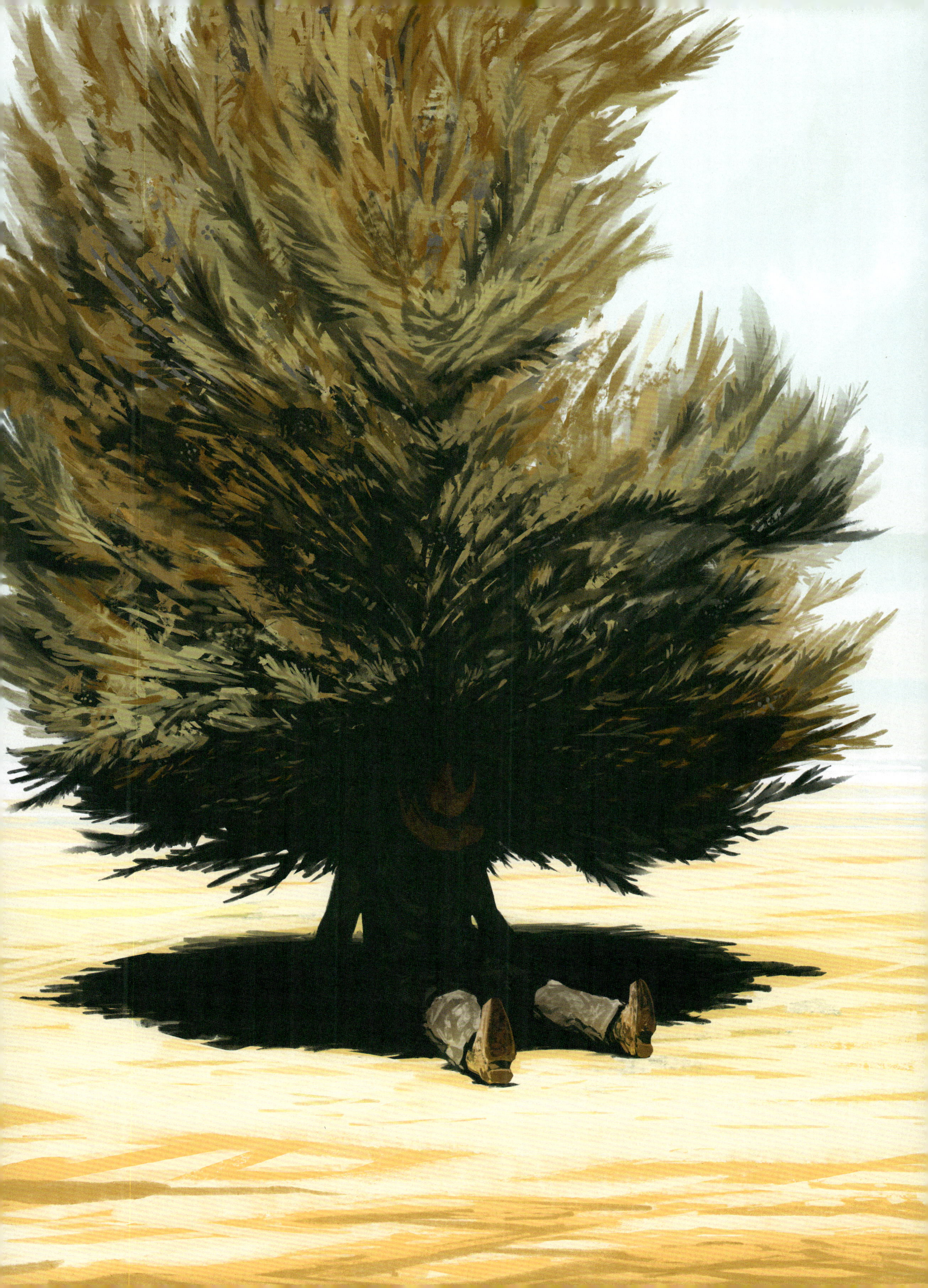

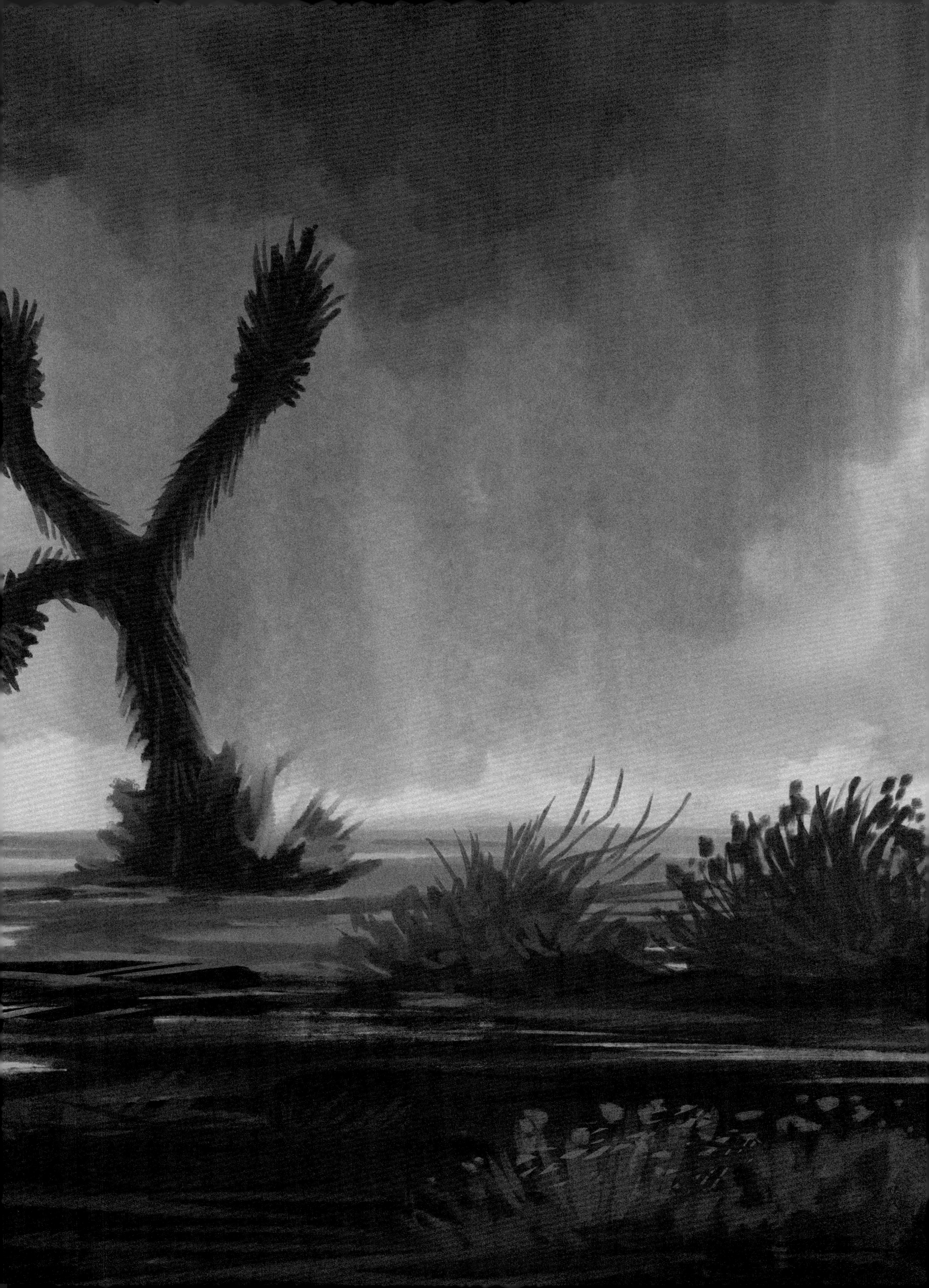

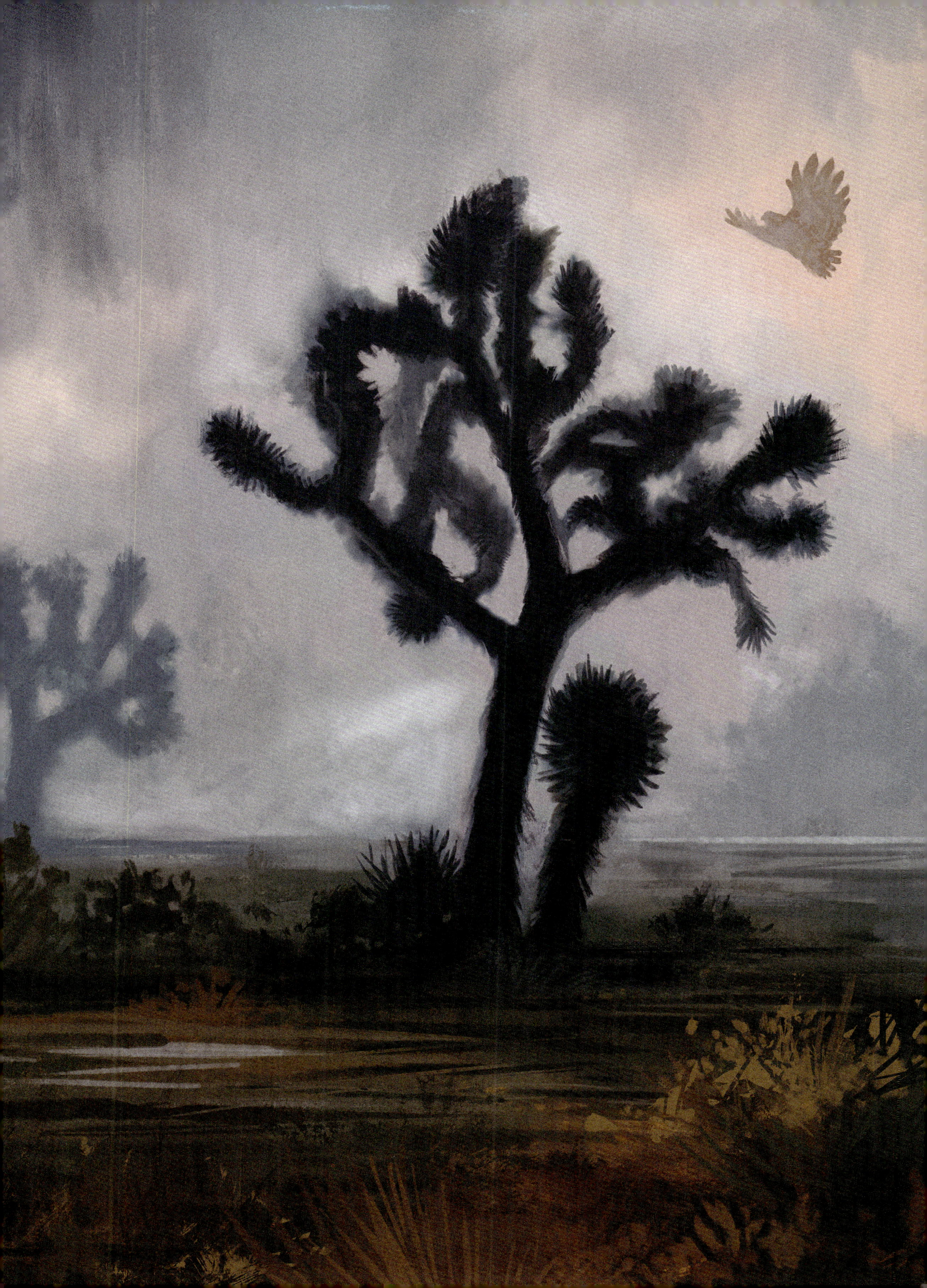

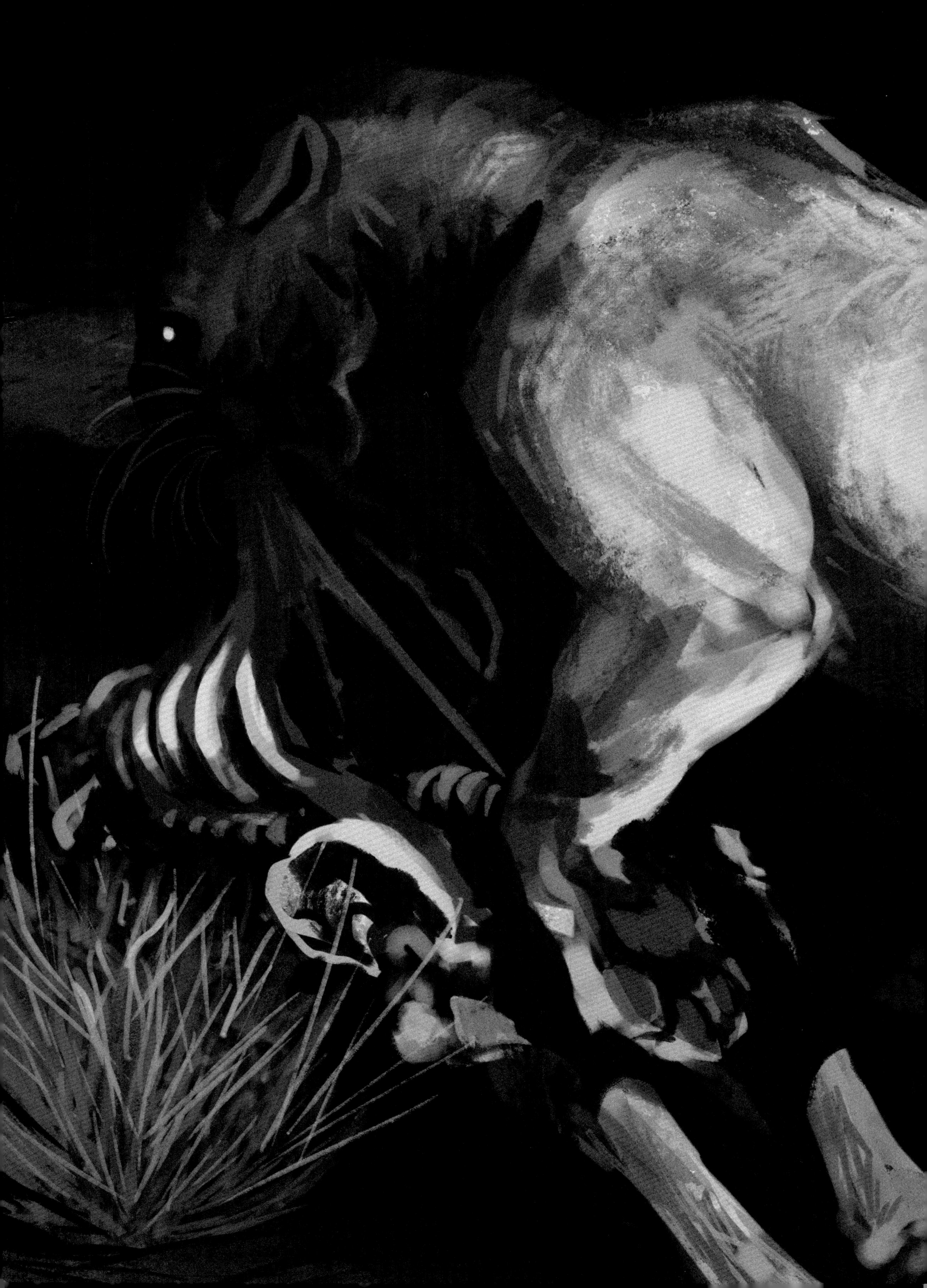

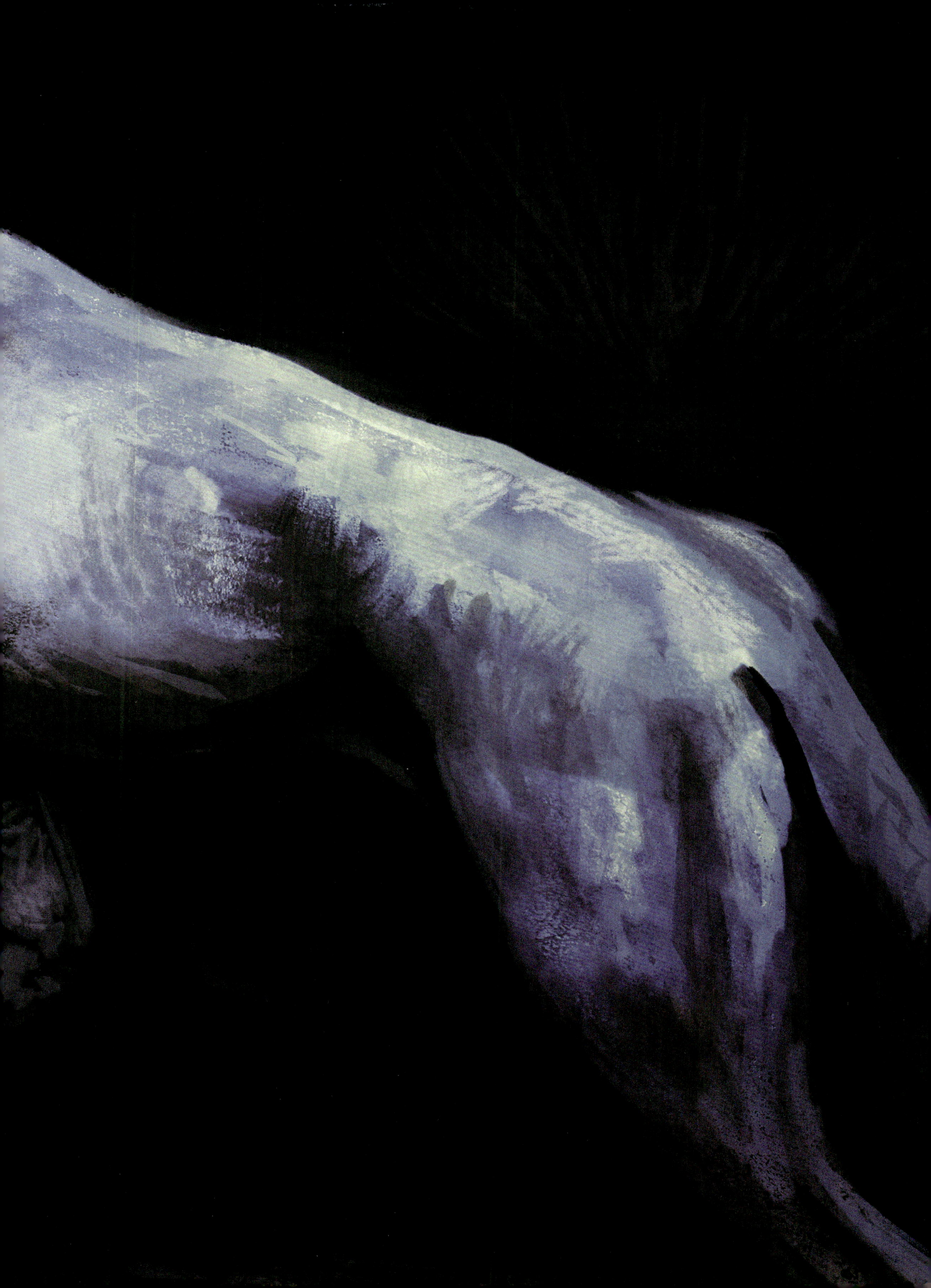

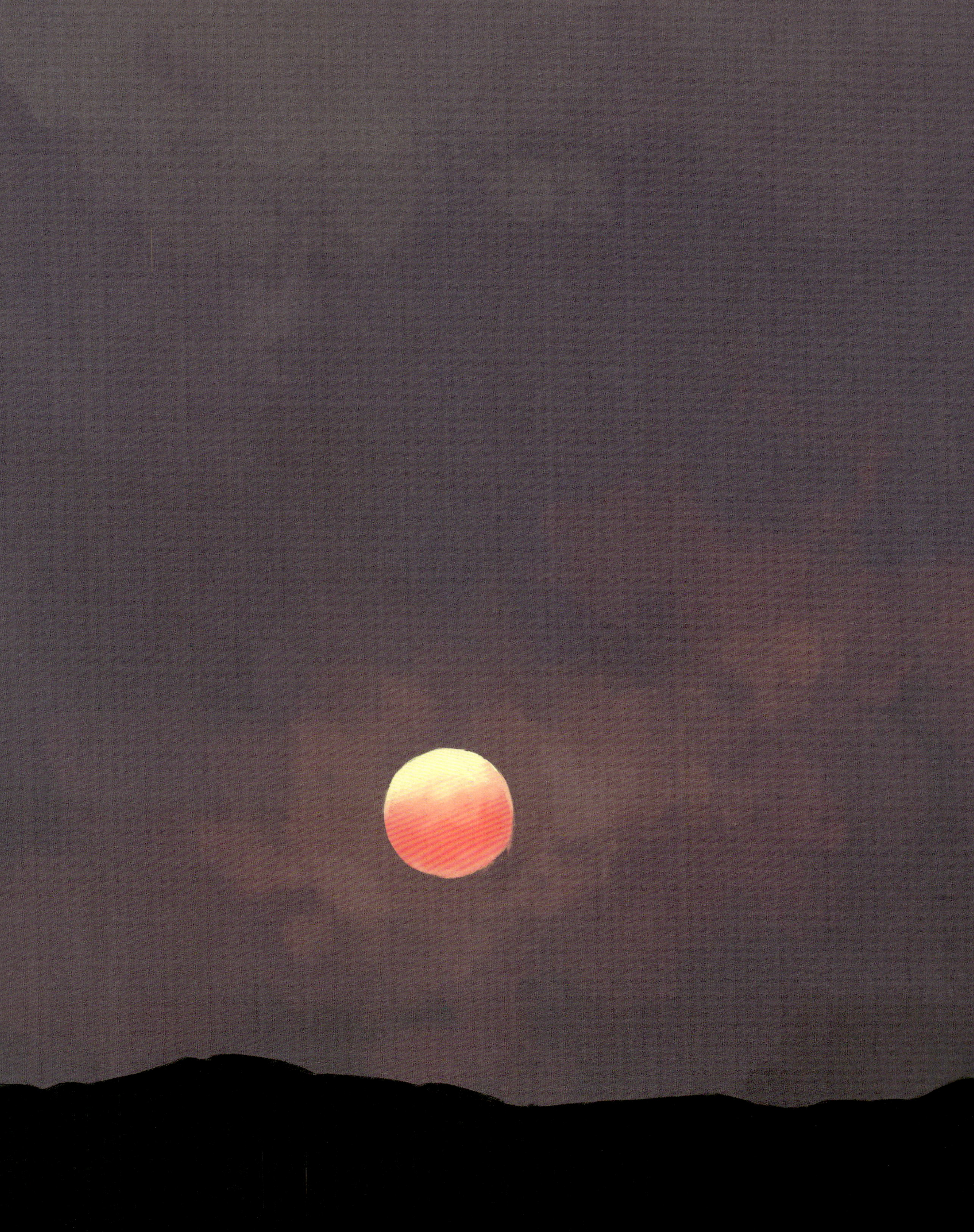

TO MY HUSBAND, JOHN.
FOR YOUR ENDLESS SUPPORT AND LOVE,
THANK YOU FOR FOLLOWING ME INTO THE DESERT.

Edited by John Fleskes
Designed by John Fleskes and Brynn Metheney
Production and book design assistance by Vicky Lien
Copyedited by Martin Timins
First Printing, October 2023
Printed in China
Asia One Printing Limited, Hong Kong
Paperback edition ISBN: 978-1-64041-076-3
Publisher signed hardcover edition ISBN: 978-1-64041-077-0
Library of Congress Control Number: 2022943280
fleskpublications.com
brynnart.com

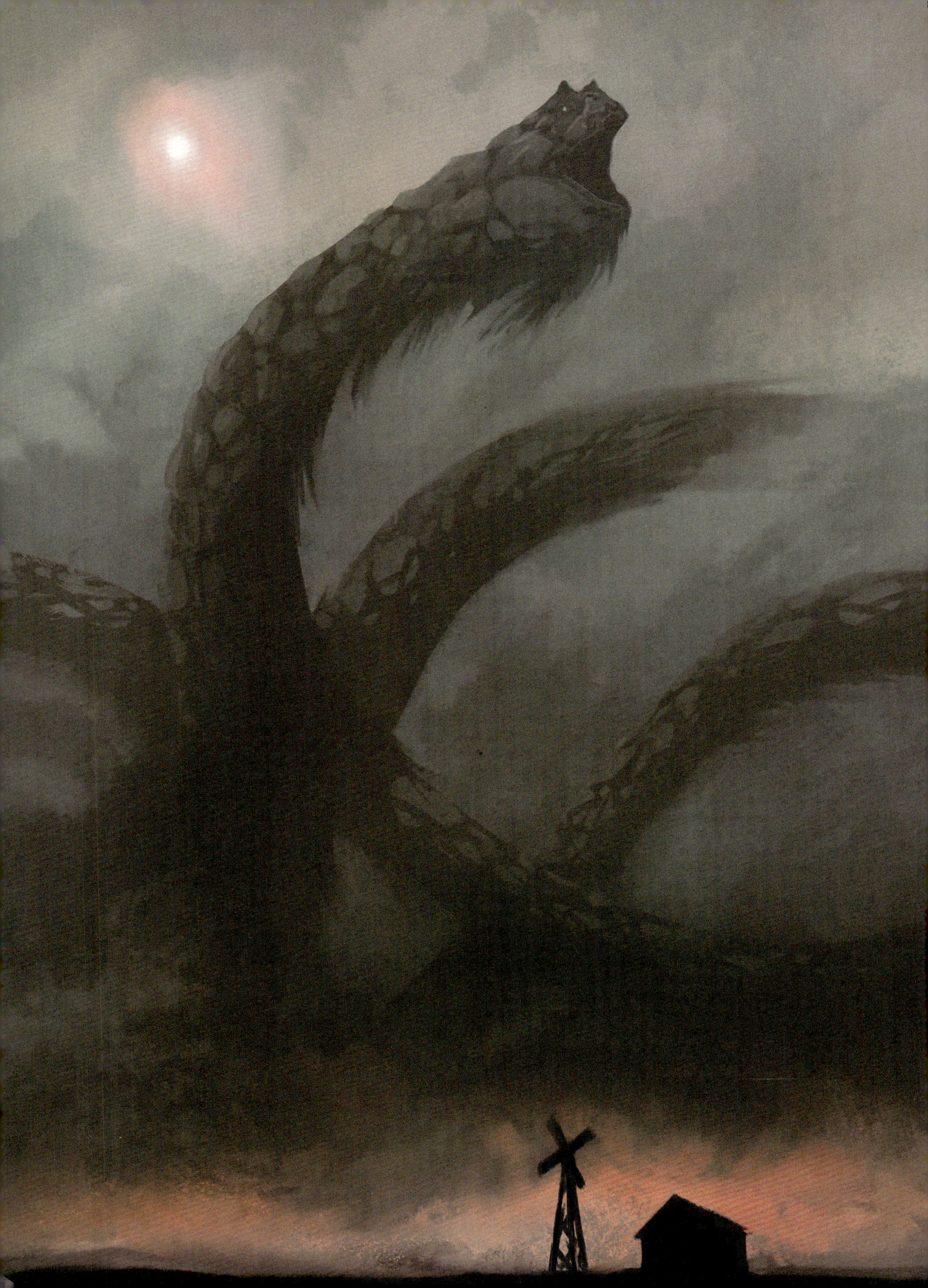